CW00548183

To Sue

# The Plant Listener

Julie C. Kilpatrick

Hope you enjoy it

Julie C. Kilpatrick

Gardenzine
Publications

First published in 2018 by
Gardenzine Publications
*www.gardenzine.co.uk*

Copyright © Julie C. Kilpatrick 2018


All rights reserved.

ISBN: 978 1 9997243 0 6

Dedicated to my mum.

## ABOUT THE AUTHOR

Julie Kilpatrick is a gold medal winning garden designer and lecturer in garden design, landscaping and horticulture. She is editor of the gardening advice website *Gardenzine* (www.gardenzine.co.uk) and, along with several guest writers, regularly contributes articles.

She tweets at @plantlistener and blogs at www.theplantlistener.com.

# CONTENTS

## Part Four: Worshipping the Sun

## Part Five: Getting Enough to Drink

## Part six: We all Have to Die Sometime

What if Plants Have Brains?

If plants had brains, they would probably have just one thing on their minds and that's sex. From the moment a plant is born until the day it dies, its whole life is governed by the need to ensure the survival of its own species. To do this, it must live long enough to produce offspring. It has to survive changes in weather conditions and it must gather and store energy and nutrients to help it grow and produce future generations. It has to endure attacks by pests and diseases and of course, if it can, it must avoid being eaten.

Their overwhelming drive to procreate makes plants the most successful group of living things on the planet. On every spare piece of ground, on mountains and hillsides, in the driest of deserts, plants grow without any help at all from us and, in many cases, in spite of us. Even in the concrete landscape of our towns and cities, you can't step outside without seeing a plant in some shape or form.

Almost every other living thing on the planet is dependent on the survival of plants for its existence and none more so than us. We are so dependent on plants we have developed a sub-conscious desire to simply look at them. We are drawn to them even if we don't know we're doing it. We want to look at them from our windows, we have them in our homes and we wouldn't consider a view to be a view without them. Plants are so commonplace in our environment they have literally become part of the furniture. We use them as decoration, placing them here and there according to *our* needs and discarding those that offend us. Vegetarians like me consider the killing and eating of plants to be more acceptable than doing the same to other living things. We are able to do these things to plants because we don't consider they have intelligence. But what if they did?

Charles Darwin believed that the tips of a plant's roots acted like the brains of one of the lower animals, receiving information from its environment, communicating that information to other parts of the plant and directing movement accordingly. Needless to say, his ideas were not accepted by leading scientists and botanists at the time. But neither, for a while, was his theory of evolution.

The argument for a lack of intelligence or the absence of brain-like functions in plants has always been very convincing. Plants are stuck in one place for their whole lives, unable to run away when danger threatens or move into a more desirable location. They cannot feel pain because pain is reserved for those living things able to react to it by avoiding it. Why would plants have an ability to sense danger? What would be the point when they can't avoid the danger anyway?

That would be a nice comforting thought for someone like me who enslaves plants to use them for my own purpose without a thought for whether or not it's what they want. As a garden designer, I list the plants required to realise the design in the same way I list the non-living materials – the gravel, paving slabs, etc. I must have destroyed thousands of perfectly good plants because they didn't fit into a particular design. I have grown plants for the sole purpose of killing and eating them. When I pull out a weed because it's competing for space with my cabbages, can I ignore the possibility it might be screaming just because I can't hear it?

The truth is I, and millions of other plants-people, can't really afford to consider the screams of plants. Even if they do indeed have any kind of brain-like function, their 'brains' would be extremely simple affairs. However, that doesn't mean we shouldn't seek to understand them. If we can 'think' like plants we can learn how to grow them more successfully. If we can empathise with them, we can make their lives so much better.

Plants are not the static, passive things we think they are. In fact, the development of time lapse photography has shown us they are far from stationary and all that movement has a purpose.

As you will discover, plants are very busy indeed. They can sense the amount of sunlight available and can move in tune with the sun's journey across the sky. They can communicate with animals and other plants. They are extremely sensitive to their environments and are aware of the presence of other plants around them. Parasitic plants can even move towards their potential victims. Plants can anticipate times of hardship and take advantage of opportunities. They even go to sleep.

Underground, their roots are travelling through the soil using movements very similar to those of worms. Each plant owns a network of roots and there are neurotransmitter-like molecules in the tips of roots similar to the neurotransmitters in human brains. This would seem to support Darwin's 'root-brain' theory. For those of us who seek to use plants – whether for food or for decoration – a knowledge and understanding of the 'plant-brain' will help us do this far more successfully than if we simply consulted a manual that told us what to do for each individual species of plant and at what time of year.

I often hear people say they kill every plant they come into contact with. Those who say this seem to think that, somehow, they're a jinx on plants, or plants in their care die just to spite them. If you fall into this category, you just don't know plants very well yet. Read on and you will find out the secret to caring for plants is being able to anticipate their reactions to certain events.

There are thousands of manuals on how to care for plants – dead-head them at this time, prune them here, sow seeds at this depth, harvest them now, and so on. But what if your manual tells you when to prune roses and not when to prune *Buddleia*? What do you do with your *Buddleia*? If you can get to know plants well enough, you will know exactly what to do with your Buddleia and any other plant you come across simply by putting yourself in its shoes.

If a plant really could think, the last thing on its mind would be to die just to spite you. In fact, as soon as you choose to grow it, your plant spends its whole life helping you do just that – all you have to do is to listen to what it's trying to tell you.

## The sensitive plant

The sensitive plant (*Mimosa pudica*) is a fascinating little oddity. It appears to have feelings because of its reaction to stimuli such as touch or heat. If you stroke a sensitive plant, its leaves droop immediately as if playing dead and the ability to do so has no doubt come about to put off potential grazers. Faced with the choice of eating a nice, healthy looking plant or a plant that suddenly appears dead and unappetising, a grazing animal will probably move on and leave the plant alone.

Not surprisingly, because it displays such a visible reaction to stimuli, the little sensitive plant has been at the centre of numerous scientific experiments over the years. This poor plant has been scratched, burnt, squashed, frozen and subjected to electric shocks. It has been exposed to very loud classical music and was even given chloroform. And yes, in case you wanted to know, chloroform does anaesthetise a sensitive plant.

Perhaps one of the most controversial experiments undertaken on sensitive plants was conducted by Dr Monica Gagliano. She dropped sensitive plants from a height onto a cushioned landing pad. The speed of the drop was enough to alarm the sensitive plants sufficiently to close their leaves, but the landing was soft enough to do no harm. After four to six drops, the plants stopped closing their leaves in response to the drop. Even more surprising, when the experiment was repeated after a month, the plants did not close their leaves. Dr Gagliano believes this is evidence her plants somehow learned that the experience of being dropped ultimately did them no harm and, therefore, did not require a defensive response. Even more remarkable is that they remembered their learning experience some time later. Many of those in the scientific community have rejected Dr Gagliano's findings, objecting to her use of the term 'learn' since plants have no recognisable brain and therefore cannot learn.

If you want to have a go at teaching a sensitive plant of your own, you can grow them relatively easily from seed, but they're native to South America so most of us would grow them as a houseplant. You can buy seeds at garden centres or online.

Soak the seeds in warm water overnight. This will trick them into believing a warm South American spring rain has descended, encouraging them to sprout. Sow the seeds at a depth of just 3mm. Once the seedlings begin to appear, keep them in a warm and sunny place, making sure the soil doesn't dry out. When the leaves of the seedlings react to touch, you can transfer each into its own individual pot.

The sensitive plant is usually grown as an annual and isn't much to look at but you can always entertain your visitors with it. However, bear in mind, all that movement is likely to exhaust the plant's energies so don't poke and prod it too much.

# Why you might talk to your plants

My love affair with plants began at the age of eight when my mother allowed me to have a patch of soil of my very own. It was a narrow, stony border at the side of the house, always dry as a bone because the house walls prevented rainwater from reaching it. In it, I planted five geraniums and I named each of them after one of the 'Charlie's Angels'. Every day, I addressed my Angels by name and had a conversation with each of them in turn because I had been told you should talk to your plants. Not surprisingly, my Charlie's Angels flourished even in this most inhospitable of borders and for a long time I believed it was because I spoke to them.

In a way it was. Talking to each plant made me pay them some attention so, mid-conversation, I would notice if the plant was in need of water, if the flower head had to be removed and so on. This is the secret behind the talking-to-your-plants-myth. It stands to reason that if you talk to something and even give it a name, you're likely to care for it more. When you talk to your plants, you also breathe carbon dioxide over them which they need to help them photosynthesise but it's not likely to be enough to make them grow any better. It's the added attention you give them that's the key.

I don't talk to my plants any more – at least not all of them - nor do I give them all names, but I do try to pay them all some attention rather than treating them as part of the furniture. In this way, I can see when and if they need my help. In the case of houseplants this attention is very important.

Imprisoned in a container within an alien environment, our houseplants are entirely dependent upon our help to survive. Incidentally, those houseplants of mine I have named get the best of care from me. Mr Pitcher the pitcher plant, Miss Venus, the flytrap and Sunny the sundew are never neglected. My carnivorous threesome gets exactly what they need, when they need it, and that's not always the case for the rest of my houseplants.

You might want to think about stroking your houseplants too. This isn't a form of therapy for you, but rather a gym session for your plants. Without passing animals or the wind to push them around, houseplants don't need to strengthen their stems as much as those that are exposed to the elements. Like anyone who doesn't exercise, they get lazy and lose some of their muscle tone. By stroking them or shoving them gently, you encourage them to toughen up.

The process of plants' reaction to touch in this way is known as thigmomorphogenesis but I prefer to call it personal training for plants.

# Part One: The Evolution of Plants

# Plants Make Landfall

Somewhere around 470 million years ago, if you had looked down on Earth from space, you would have seen a planet relatively similar to how it looks from space today. More than 4 billion years after it first came into being, Earth had continents and oceans, even though we may not have recognised the shapes of the continents. England and Wales were in the Southern Hemisphere while Scotland and Northern Ireland basked in a tropical climate near the equator. Much of North America was with Scotland on the continent of Laurentia while parts of the Eastern seaboard were on the micro continent of Avalonia, alongside England.

The continents were probably bare. Some say there was nothing living on land at that time; others say there might have been a small number plants and fungi since millions of year before. One thing is for certain, the great continents back then were all but bare rock and nothing like the rich, lush, green landscape we enjoy today.

In the oceans, however, the picture was altogether different. The Panthalassic, Paleo-tethys and Iapetus Oceans separating the continents were teeming with life. Diversification in the sea had begun many years previously and there were crustaceans, molluscs and weird, jawless fish.

Photosynthesising organisms had steadily increased the amount of oxygen and food available in the seas, causing an explosion of diversity. Amongst the photosynthesising organisms in the seas were algae, givers of life in the oceans and about to do the same for land.

The first algae to make landfall got there by accident, washed in by tides or left behind when the unstable climate flooded continents and then retreated. Accustomed as they were to a life surrounded by water, they found themselves in a hostile world, completely unequipped to cope with the desiccating effect of this new place. Those landing in shallow rock pools fared reasonably well, but any unlucky enough to be out of water for any length of time, perished.

Beyond the rock pools was a vast stretch of rocky land. It was a huge area just waiting to be conquered but, to lay claim to this new environment, the algae had to adapt. Surviving in the shallow rock pools was no picnic compared to life in the deep oceans. The algae were naked and exposed and they had to find a way to cover up. They developed a cuticle – a skin-like layer that protected them from the drying effects of life on land. Pores in the cuticle helped them to absorb carbon dioxide, water and oxygen.

It is generally accepted that the liverworts evolved from green algae and they were amongst the first land plants to adapt to life beyond the oceans. Liverworts are simple, moss-like plants with flattened leaves. They absorb moisture through all parts of the leaves but are incapable of transporting water throughout the plant so they must be in constant contact with it.

The early liverworts lived on rocks close to the water's edge. Though they lived close to water and depended entirely on it, there were times when they were left high and dry and at the mercy of the drying effects of the sun. So liverworts developed a feast or famine approach to life, an on-off switch if you like. When water was available, they went to work, collecting energy from the sun to help them grow. When water was unavailable and they dried out, they could switch to a dormant state and wait a while until water re-invigorated them again.

Liverworts also relied on contact with water for reproduction. Male and female parts developed on different plant bodies and liverwort sperm were required to swim across a film of water to find a female. If a swimming sperm made contact with an egg, the result was the formation of a sporophyte which grew on the parent plant and contained spores – the next generations. Spores were then released during dry periods and carried on the wind to new sites, but these spores had to be lucky enough to land on moist ground before they could complete the cycle and turn into brand new plants.

The adventurous liverworts had set out beyond the relative safety of the oceans in search of new lands, but they didn't really get very far. They could only stare over towards the undiscovered horizon with longing. Trapped in areas extremely close to water, they huddled together in groups in an attempt to keep humidity levels within their colonies as high as possible.

Also relying on safety in numbers were the mosses. Mosses are thought to have evolved from liverworts and they, like the liverworts, required a constant supply of water. While neither the mosses, nor the liverworts had roots as we know them today, they did eventually develop little root-like structures known as rhizoids which could anchor them to the land. The rhizoids of mosses were more efficient than those of the liverworts so the mosses could cling more easily to the sides of rocks.

You can, of course, still see mosses and liverworts today and they continue to occupy the spaces they would have preferred all those years ago. Take a walk through any damp woodland and you'll see all manner of mosses and liverworts. If you're very lucky, you may even catch sight of their cousins, the hornworts, which look a little like liverworts but are characterised by their hornlike sporophytes. Look for liverworts, with their underdeveloped rhizoids near the ground, pressing themselves up against any available moisture. Mosses have slightly more complex leaves and they can grow a little bit taller because their rhizoids are capable of absorbing some amount of water and nutrients. Hornworts can be found all over the world but they prefer warm, tropical climates.

Today, the liverworts, hornworts and mosses still prefer to live in large groups, just as they did all those millions of years ago. They continue to rely on water for reproduction and to retain the ability to survive periods of drought by entering a dormant state.

Most scientists agree that the liverworts and later the mosses are the descendants of algae from the sea, but this adaptation from sea to land came about very slowly since the algae had to completely alter their original structures. However, it was not the only way algae were able to colonise the land. There was another way for some to survive in a dry atmosphere and it required the help of an entirely different species altogether.

Sometime after leaving the sea, certain species of algae happened to bump into certain species of fungi and together they formed friendships that still exist today. The fungi needed to find photosynthesising organisms to use as a food source and the algae were in need of a protector. Lichens are not quite plants and not quite fungi. They are, instead, the result of a weirdly symbiotic (mutually co-operative) relationship between algae and fungi. Somehow, a fungus discovered a lonely alga and surrounded it with the intention of consuming its sugars. However, instead of simply eating the alga and moving on, the fungus chose to stay with it, protecting it from death by desiccation and, in return, the alga gave the fungus the life-giving energy it needed to survive.

This strange marriage of two very different species still persists today. Like any singleton, each alga and each fungus applies certain sets of criteria when choosing life partners. Once they have found a suitable partner the two parties become so closely interwoven that two become one and the lichen is given a species name of its own. There are thought to be around 30,000 species of lichen in the world.

Lichens can grow just about anywhere, even in places where no other plants can. They remain a very important pioneer species, colonising bare rocks and even growing on our walls and pavements. The next time you see what looks like chewing gum on the ground, it may not be chewing gum at all but lichen and it may very well be much older than you. They grow very slowly, some only a millimetre a year and can live for thousands of years. They occupy about 8 per cent of the land surface of the Earth and come in all sorts of shapes, sizes and colours. Perhaps one of the best places to see lichen is hanging around graveyards where they keep silent company with our dead, living on old gravestones.

We live our lives in the blink of a lichen's eye. Long after we are gone, their enduring marriage will no doubt pave the way for the arrival of more complex species because, when they eventually do die, their remains form pockets of decaying matter that will feed the seedlings of more complex plants.

In the early years, lichens, mosses, hornworts and liverworts enabled animals to follow them out of the sea. Protected from moisture loss under the shade of the plant pioneers, these tiny creatures fed off land plants just as their ancestors had fed off algae in the sea. When they and their associated plants died, the organic matter they left behind combined with pieces of weathered rock to form the first soils and it was the formation of soil on land that allowed plants to take a huge leap forward in their bid to colonise Earth.

## Mosses under threat

Sphagnum mosses, otherwise known as peat mosses, have adapted to their own particular water requirements by developing cells which can hold up to twenty times their own weight in water. When the soil dries out, sphagnum can call upon its water reserves to keep it going. Like most mosses, they grow together in large numbers and, since they live and die in moisture-rich, oxygen-poor environments, the remains of all the dead plants in those areas form tightly packed layers beneath the live crops. This layer of dead material – known as peat or peat moss – is much prized by gardeners as a soil conditioner. Because of its ability to hold water, we also use sphagnum moss itself to line hanging baskets.

For most plants, it is usually beneficial to be of some use to humans. They are protected by us and distributed throughout the world, adding to the species' chances of survival, but the same cannot be said for sphagnum. Sphagnum moss peat bogs are unique and home to a number of different species of living things that could not survive without the mosses. It takes thousands of years to form peat and, once gone, it takes many years to get it back. The mining of peat by hand for use as a fuel source was relatively sustainable because it only removed small bits of peat at a time and allowed the habitat to recover. However, for the horticultural market, peat moss is often mined mechanically. For this to happen, the peat bogs must first be drained resulting in the loss of habitats which can never be replaced.

On top of that, peat bogs are considerable carbon sinks. Worldwide, it is estimated peat stores around two times as much carbon as forests but, when peat bogs are damaged, they begin to degrade and release carbon back into the atmosphere. With around 15% of the world's peat bogs, the UK government was sufficiently alarmed by the threat posed by horticulturalists that it paid over £17 million to a large manufacturer of potting composts in order to persuade them to stop extracting peat from certain sites.

There is no longer any excuse for us to use peat in large quantities since there are now so many alternatives. With the lure of large pay-offs to discontinue mining peat and the threat of a ban on the use of composts containing peat, the compost manufacturers have been hard at work searching for alternatives that will perform as well, or better than, peat. Peat-free soil conditioners are readily available and you can easily locate peat-free composts in all the usual shops. You can also buy a number of alternative liners for hanging baskets, for example, those made of jute or cardboard so you don't have to harvest live moss either.

Like all wild plants, sphagnum mosses are afforded protection by law and it is an offence to collect sphagnum moss and other wild plants without the permission of the landowner.

Generally, the collection of mosses from the wild is only allowed in small quantities and from plantations of conifers that are shortly to be felled. To avoid the risk of prosecution, I would advise you not to collect wild moss at all and, if you absolutely must buy it, look for evidence on the packaging to tell you the moss has been collected responsibly.

## Are you going to let a basic plant like moss get the better of your lawn?

One of the more common problems I get asked about is how to control moss in a lawn. What most people are looking for is a quick fix – something they can add to the lawn to kill the moss - but this is usually only a temporary solution. To understand how to control moss effectively, you need to apply what you know about mosses to solving the problem.

We already know mosses do not have roots to speak of. Because of their place in evolution, they are part of a group of plants known as lower plants. On the other hand, our lawn grasses are highly developed. They have roots, a vascular system and complex reproductive habits mosses can only dream about. If your higher lawn grasses can't compete with the lowly mosses, then it usually means there's something wrong with the grasses. In other words, it's not so much that the moss is strong, but more likely the grass is weak.

Grasses have well-developed root systems and look for all their nutrients in the soil. Mosses can survive in poor soil conditions because they don't rely on soil for their nutrients or water. To ensure your grass can out-compete moss, give your lawn a good feed. You can use an over-the-counter lawn feed in spring (you can even get organic ones) or you can use sieved home compost spread over the lawn as topdressing in autumn. If you want a lawn that is your pride and joy, why not do both?

To spread lawn fertiliser, make sure you cover the lawn evenly and thinly. I prefer to spread by hand since sometimes, when you go over a bump with a spreader, it dumps too much fertiliser in one place which can burn the lawn and cause it to go black. Go over the lawn in one direction, spreading the powder as you go, then change directions and go the other way. Keep in mind these fertilisers are fairly instant and any bits you don't catch with the powder will not be as green as the bits you do. One of the first jobs I ever did was to fertilise a client's lawn which I have to admit, I completed in a rush. I was called back to visit the client a week later to discover she had a striped lawn – green where I'd caught it with the fertiliser spreader and yellow-green where I hadn't.

After you've spread the fertiliser, wait for rain and, if it doesn't happen within a day or so, give the lawn a good soak to ensure the powder is dissolved and washed down into the soil. You'll soon find that your lawn will start growing away nice and strong and begin to out-compete the moss.

To apply topdressing, you can buy a bag from the garden centre or make your own using a mix of finely sieved homemade compost and sand. You could try adding some lawn seed to the mix to help thicken up the grass if it needs it. Spike the lawn all over with a garden fork to promote good aeration – something the grass roots will love –then spread the mix over the lawn and work it in using a brush or the back of a rake. Make sure the blades of grass are still visible and the topdressing mix is worked all the way to the bottom.

Once you've improved the fertility of the soil, raise the height of the blades on your lawn mower. Since mosses don't have the ability to grow very tall, the grass will be able to compete for sunlight more effectively if it can tower over the moss. When it gets to midsummer and the weather is dry, lower the blades and cut the grass short. This will let the sun dry out the moss, weakening it even more. Cutting lawns too short at the wrong time is a common cause of excess moss in lawns.

Mosses love damp, shady conditions. If you have a particularly shady lawn, do what you can to reduce shade by cutting back shrubs and removing one or two branches from trees to open out the canopy. Buy grass seeds of grass species especially adapted for shade, mix them with fine compost and spread the mix over the lawn.

Finally, don't be tempted to scarify the lawn while the moss is still green. Scarifying is done by vigorously going over the lawn with a lawn rake or powered scarifier and many people think the purpose of scarifying is primarily to remove moss. In fact, though it does pull out a fair amount of moss, the main reason for scarifying is to remove dead grass roots near the top of the soil. Remember, lawn moss isn't like ordinary weeds. Since moss doesn't have a root system, it really does no good at all to 'uproot' it and you'll only encourage the moss to spread by raking it around. Mosses reproduce in two stages. First, the egg is fertilised by the swimming male sperm and for this they need water. Next, the spores are produced which will eventually form the new plants. These spores can survive without water and hang around until it rains so raking out live moss spreads the spores and, as soon as the weather gets wetter in autumn, you'll have more moss than ever.

For really bad infestations of moss, apply a moss killer or a combined feed and moss killer. Wait until the moss turns completely black and then you can scarify. After that, re-seed the lawn to encourage it to thicken up and fill in any bare patches. Following this treatment, you must be vigilant. Keep the lawn well fed and the height of the lawn above the normal growing height of a moss plant for most of the year. Make this cutting and feeding regime a habit. As soon as you allow your lawn to revert back to the kind of conditions mosses love, your moss will return.

While not a quick fix, the long-term answer to controlling moss is not about continually attacking the moss but about looking after the grasses as well as you can. A healthy, thick lawn is a lot less likely to support moss.

# Putting Down Roots

When mosses, liverworts and lichens first stepped onto land, the levels of carbon dioxide in the atmosphere are thought to have been around twenty-two times the levels of today. The weather was warmer and wetter with frequent storms and floods. This moisture-rich atmosphere was, of course, perfectly suited to mosses and liverworts and it was a smart move to leave the sea when they did because, not long after, there was a mass extinction event, occurring mainly in the oceans.

No one knows exactly what caused this, but some say the arrival of primitive land plants was a contributing factor. As they went about their daily lives, by secreting acid, they may have caused weathering of the rocks they lived on. Pieces of weathered rock were washed into the sea and the minerals fed marine life, allowing it to reproduce rapidly. Oceanic life increased in volume and took up all the available oxygen, so that even the photosynthesising algae could not replace the oxygen quickly enough. Too much competition for oxygen led to a large number of deaths and, without oxygen to help break down the dead bodies and release the carbon, the carbon was locked up and could not be released back into the atmosphere as carbon dioxide.

On land, the early plants were using up carbon dioxide as well. As carbon dioxide levels in the atmosphere dropped, the Earth began to cool and, as the temperature dropped, many species used to warmer waters were lost altogether. Millions of years later, we would discover the carbon locked up during this mass extinction event and release it back into the atmosphere by burning oil and gas, causing carbon dioxide levels to rise again.

While their oceanic neighbours might have suffered, some land plants emerged relatively unscathed. The polar ice caps formed, but that didn't bother the plants inhabiting the warmer parts of the globe. The drop in temperature and in carbon dioxide levels led to a more stable climate without so many unpredictable storms and floods, but it still wasn't nearly as stable as the one we know today. While the mosses may have done us a favour all those millions of years ago, the job of creating, and then sustaining, the atmosphere we enjoy today began with the arrival of the vascular plants.

When the rootless plants and their associated animals died, layers of soil built up on the once-barren rocks. Plants now had something to cling onto. Enriched by decaying organic matter and minerals from weathered rock, the new soil was not only a source of nutrients but could hold onto rainwater. If plants could only find a way to harvest this store of water and nutrients, they could travel further inland. Mosses and liverworts carried on living in wet environments close to water, but an adventurous few ditched their thin stringy rhizoids in favour of thicker, stronger anchors.

The evolution of roots meant that, not only could plants hold fast to the land, they could use their roots to suck up water and nutrients from the soil. This set them free from their watery prisons and allowed them to seek out new territory. They could finally travel across the land, creating soil as they lived and died in each newly colonised site. The deeper the soil, the bigger the store of water and nutrients, and the deeper and wider the roots could spread in order to anchor the plants. They each fought for every bit of sunlight available, developing stronger roots, taller stems and ever more diverse leaves to capture the sun.

This was the age of the club mosses, ferns and horsetails - the first vascular plants. They were able to transport water and nutrients against gravity, into the stem and all the green parts of the plant. Like mosses and liverworts, they reproduced by swimming sperm so they still had to maintain humidity and continued to rely on safety in numbers. They lived side by side, in huge forests of weird giant ferns and strange-looking, massive horsetails.

Having found this new way to transport water and nutrients, they could experiment with shape and size, with some growing up to 35 metres tall so that smaller plants were able to occupy the understorey beneath a protected rainforest canopy. The mosses, liverworts and their associated animals could travel along beneath the humid umbrella of the vascular plants. This new generation of plants did something more than realise their own wanderlust, they allowed the others to come along for the ride.

The land they occupied was still frequently flooded and, like their sea-dwelling ancestors before them, when the great club mosses and horsetails died, there was very little oxygen to speed up their decay. Their remains formed first peat and then coal, plucking more carbon dioxide from the air without returning it and altering the atmosphere in the process.

The carboniferous (coal-bearing) period lasted around 60 million years and, during that time, there was another extinction event, but this time the land plants and animals bore the brunt. The Carboniferous Rainforest Collapse occurred due to a cooling in the atmosphere followed by a period of global warming. Other factors may also have contributed to the change in climate, but, in any event, this mass collapse of the rainforests led to the extinction of plants and animals alike and the fragmentation of the rainforests. The surviving plants were marooned in little pockets of swampy land. They had survived because they had been lucky enough to find themselves in sites where low-lying land had captured enough water, but they were now trapped there, surrounded by drier land, no longer living in a vast forest with the high humidity and protection they had enjoyed before. They had found a way to utilise stored water in the soil, but, to colonise all parts of the land, they would have to ditch their reliance on water for reproduction.

## Horsetail – the living fossil

Most gardeners hate horsetail (*Equisetum arvense*) because, once established, it is notoriously difficult to eradicate. With its tough outer body, it is able to shrug off any attempt to treat it with weed killer. It can re-grow from a small piece of root and the roots are annoyingly thin, so you never seem to find every bit. Since its ancestors were around more than 300 million years ago and the genus survived a mass extinction event, you might understand why it doesn't appear to be too bothered by our puny efforts to destroy it.

The horsetail species of today are not nearly as tall as their carboniferous ancestors but, like them, they do seem to rely on safety in numbers. They still insist on reproducing from spores, so they require a moist soil or at least a moisture-retentive soil. Single spore-bearing stems appear before the leaves and often go unnoticed. By the time the green parts of the plant are fully visible, the damage is already done and the next generation is waiting in the wings. Digging them up is not easy since the roots can be very deep and extensive.

Horsetails are rich in silica – which is why weed killers have little effect on them – and they were once used as scouring pads so if you have horsetail in your garden, you can at least use them to clean your dishes. A good organic way to get rid of this prehistoric relic is to hoe off the top parts of the plants as soon as they appear. If you do this constantly for a few years, you'll prevent them from photosynthesising and weaken the plants. If you feel you need to use chemicals, it is sometimes recommended you first crush the leaves to allow herbicides to penetrate the plants' protective silica and then apply your weed killer but this method also requires several treatments over many years. Not being a lover of chemical control, I'd always recommend the traditional, manual approach.

If you don't hate horsetails enough, you might be interested to know they are credited with the discovery of logarithms. Scottish mathematician John Napier is said to have been inspired by the way horsetails grow in decreasing, uniform segments that appear to telescope into one another.

## How to grow a prehistoric monster

If you want to grow something from a prehistoric age that isn't an annoying weed, you could try growing a tree fern. Forests of tree ferns existed millions of years ago and pre-dated the arrival of seed-bearing plants. In a garden, one tree fern specimen can look pretty impressive. While it won't reach the heights attained by its ancient ancestors, it can grow reasonably tall. The best species to go for is *Dicksonia antartica* which, as its name suggests, is able to withstand lower temperatures than most. It comes mainly from Tasmania and the ones on sale are harvested under license.

Tree ferns are expensive. As a general rule, the more expensive a plant is, the slower it is likely to grow and this is true of *Dicksonia* which has a growth rate of between 1 and 10 cm per year. Because of this, you will pay by the metre for your fern and how much you spend depends on how long you are prepared to wait for it to grow.

Having shelled out a small fortune for your plant, it makes sense to give it everything it needs for a long and happy life. Plant your fern in a sheltered spot and add plenty of organic matter to the soil around it. After planting, spread a good layer of leaf mould or peat-free compost around the trunk to conserve moisture. On dry days, soak the leaves to bring up the humidity. Like its ancestors, your tree fern won't like to be alone, so planting smaller ferns or other plants at the foot of it will help bring up the humidity to levels the tree fern prefers.

People often think tree ferns are rootless but, being the descendants of the first root-bearing plants, this is not true. The rootless myth comes about because they are often reduced to trunks without leaves or roots for ease of transportation. It is possible to do this because, once in contact with soil, the plant will quickly re-grow a root system. Don't assume that because you receive it without roots you can get away with giving it a shallow soil. Tree ferns can develop extensive root systems and will devour any organic matter in the soil so you need to provide your tree fern with a rich soil and plenty of depth.

It may be more winter hardy than most but bear in mind that *Dicksonia antartica* doesn't actually come from the Antarctic. It is fairly hardy in mild, coastal areas, but outside of these areas, it needs winter protection. The crowns of tree ferns are very vulnerable to prolonged periods of frost, so when the fronds turn brown in autumn, don't be tempted to cut them off. Instead, bend them over the crown and wrap the whole thing up in horticultural fleece. The fleece allows moisture and light to penetrate but protects the stem and crown from frost damage. If the thought of looking at a stump covered in fleece for several months doesn't thrill you, then make your tree fern a little jacket of straw, but be sure to wrap both the stem and the crown for best protection. In April, remove the protection and wait for the new fronds to appear. If there is a late frost, the new growth might get set back a bit but it should recover.

# The Seed Plants Arrive

Around the time of the Carboniferous
Rainforest Collapse, the climate switched from
mainly warm and wet, to cooler and drier. With so
much carbon locked up underneath the collapsed
rainforests, there was more oxygen in the air and,
following a lightning strike, this oxygen could feed
a forest fire, allowing it to spread for miles. Most of
the land masses on Earth had come together to
form a mass super-continent named Pangaea. Sea
levels dropped and, with only the outer edges of
Pangaea in contact with water, this massive
continent had large amounts of dry land in its
centre. For the land plants, clearly something had
to be done about the swimming sperm problem
and, with the climate in a state of flux, there had to
be a way for them to reproduce so their offspring
might survive any seasonal changes, forest fires
and even mass extinctions.

The first plants to solve this problem were
the gymnosperms. Their sperms were aviators, not
swimmers, travelling in light and dry pollen. In a
less humid environment, these plants could
reproduce much more efficiently than their
ancestors. Once the pollen made contact with a

female, it produced a tube to allow safe passage of the sperm to the eggs without the need to swim across a film of water. With a tube to guide them, the sperm could reach their targets far easier than they could by swimming around aimlessly in a pool of water.

Alongside the ability to produce pollen, the gymnosperms evolved the first seeds. Seeds were a genetic mix of both parents alongside a little packet of food and, more importantly, they could delay growth until the conditions were right. They could wait around during dry or cold periods and, if they got buried under the soil, they could survive forest fires and other life-threatening events.

The first seeds were naked. In other words, they developed without the protection of an ovary. Despite their lack of a protective ovary, these little potential lives were able to survive more changeable conditions than the young of the mosses and ferns. This evolutionary leap was so successful that the gymnosperms were able to recover from the greatest mass extinction event of all.

Sometime around 250 million years ago, over 90 per cent of life on Earth became extinct. This extinction event was so catastrophic, it is known as 'The Great Dying'. There was no single cause, but rather a combination of factors just like previous extinction events. Though many of the early gymnosperms were lost, their seeds were able to survive and the group itself eventually recovered to achieve full dominancy of the land.

The enhanced ability to reproduce meant the gymnosperms could diversify and form new species of plants with very different characteristics in terms of growth habit and structure. To achieve great heights, the gymnosperms had woody stems that could hold fast in the face of wind, and anything else which might try to push them over, and it's just as well they did because the gymnosperms weren't the only ancestors of life previously confined to the oceans. After the Great Dying, gymnosperms were the dominant land plants of the Jurassic period and dinosaurs were the dominant land animals. You had to be made of strong stuff to stand tall in the face of these giant reptiles.

Some say it was the dinosaurs that ruled the land but really it was the gymnosperms. They towered over the landscape and released their seeds to be carried on the wind to new territories – territories on which the mosses, liverworts, ferns and horsetails could not easily survive. They had solved the problem of the swimming sperm and had found a way to give their offspring a far greater chance of success. Leaving the mosses, liverworts, horsetails and ferns in their watery prisons, they had discovered a new freedom and had even survived 'The Great Dying'. Surely no other group of plants could do better than that?

Unfortunately, to achieve lasting greatness, you have to be versatile and the gymnosperms simply weren't versatile enough to remain dominant for long. They eventually lost their place at the top of the plant hierarchy and had to give up their crowns to the most sophisticated plants of all. However, like the liverworts, mosses, ferns and horsetails, they still exist today - split between four groups – the conifers, ginkgo, cycads and gnetales.

So what kind of king can steal the crown of such a sophisticated ruler? Well, you don't necessarily have to be bigger to be better, but you do have to be smarter and the smartest kings of all make use of their subjects. The plants that reached the highest level of sophistication were able to do just that.

## The last of its kind

When an atomic bomb was dropped on Hiroshima towards the end of World War 2, six trees surprised everyone by surviving the blast. They were *Ginkgo biloba* trees, gymnosperm relics of the Early Jurassic period and their remarkable ability to survive has earned them a reputation as the bearers of hope. One of the trees grew on the site of a temple which was destroyed by the blast and, rather than cut down the tree to build a new temple, the new temple's staircase was built around the tree. The tree bears the engraving "No more Hiroshima" and has become an international symbol.

This is not the first time the *Ginkgo* has survived against the odds. Once widespread, *Ginkgos* went into rapid decline until it was thought the whole genus was extinct. Luckily for *Ginkgo biloba*, Chinese monks continued to cultivate it in their remote monasteries and it was eventually re-discovered. Perhaps because *Ginkgo biloba* is the only survivor of its genus, it is relatively free of pests and diseases, and this, coupled with its seemingly indestructible nature, has popularised it as a street tree, tolerant of a wide range of conditions and pollutants. However, this plant that evolved to stand up to dinosaurs can grow to a height and spread of 21 metres so it needs a great deal of space and isn't particularly suitable for growing alongside our busy roads.

*Ginkgo biloba* provides us with a good argument for the need to preserve as many plant species as we can. It is highly prized in herbal medicine and is said to help improve memory and even slow the progress of dementia. Researchers are looking closely at this plant and though the evidence is not yet solid, there is some hope it could be of use to us. Had the Chinese monks not offered it sanctuary, we might not have the potential to benefit from its curative properties.

# The largest living gymnosperm on Earth

Top of my list of beings I want to meet is General Sherman. Of course, the human version of General Sherman died some time ago, but his namesake still exists as the largest living single stemmed tree by volume. The only way to truly be sure of General Sherman's age is to wait until he dies and count the rings on the main stem, but at around 2,500 years old (give or take 500 years or so) he is a young pensioner. He is a Giant Sequoia - *Sequoiadendron giganteum* - and redwood ancestors of his would have been around hundreds of millions of years ago and they can live for as much as 3000 years.

General Sherman lives at Sequoia National Park in California. He was born towards the end of the Iron Age and, during his lifetime, he has seen humans go from fashioning basic tools, to creating vessels that could reach into space. Like other redwoods, he has a deceptively soft and spongy bark and it is possible to punch his bark without too much damage to your hand. Native Americans might have used him to practice their fighting skills and he would certainly have been witness to their persecution at the hands of the white man.

The diameter of his trunk is currently just over 11 metres and his circumference at ground level is over 30 metres. His mass is estimated at around 1,900 tons.

Luckily for General Sherman, he is not one of the giant sequoias whose trunks were cut out to create tunnels for humans to drive through. The last giant sequoia 'tunnel tree' fell during a storm in January 2017. The Pioneer Cabin Tree was thought to be over a thousand years old which means it died young. Though the Pioneer Cabin did a grand job at bringing tourists to the area in which it lived, I think it's reasonable to assume that carving a tunnel into its trunk might have contributed to its early death.

## Conifers in the garden

It can be very tempting to use conifers to retain interest in our gardens throughout the year. Most conifers are evergreen and these days there are so many different varieties, you can get them in all sorts of colours from golden yellow through to silver-blue. If you want to grow conifers small enough for the average sized garden, be aware that so-called dwarf conifers are not always what they seem. Sometimes they are classed as dwarf because they grow very slowly and some conifers sold as dwarfs can reach annoying heights.

Before you buy conifers, check in a good plant encyclopaedia to see what the eventual height really is because, unfortunately, it's not that easy to prune conifers if they get too big. As a general rule, you should assume your conifer will not recover if you cut into brown wood and you must always trim the green growth only. If you cut the top off a pyramid-shaped conifer, for example, it may not grow back into its former shape. When planting a border with conifers, do your research and try to choose true dwarf types. Use prostrate conifers that will grow horizontally along the ground, globe types with rounder growth and conical or pyramidal types to give you a bit of height.

If your conifer turns brown on just one side of the plant, it is probably suffering from wind burn, a condition caused by the drying effects of cold winds in winter. The leaves on the windward side of the plant turn brown in response to water loss. The only solution to wind burn is to move your conifer to a more sheltered spot or to put some sort of protection on the windward side. Unfortunately, only a few varieties of conifer are able to recover from wind burn so, if wind burn has occurred, you may have to throw the affected plant away and replace it with another under the protection of a windbreak.

For a conifer hedge, try Yew *(Taxus baccata)* as it is the most forgiving if you cut it back hard. Leyland Cypress *(Cupressus x leylandii)* is often used hedging plant because it is quick growing but it doesn't tolerate cutting into the old wood quite so well, so cut your Leylandi hedge little and often and always cut into green growth.

Conifers are excellent low-maintenance plants and pretty undemanding once established. Look after your newly-planted conifer by supplying it with adequate amounts of water in its first year and it will reward you by growing away quite happily for the remainder of its life. It won't do any harm to give it a small amount of general-purpose fertiliser each year, usually in late winter to prepare it for a burst of growth activity in spring. Better still, mulch around the base with homemade compost in autumn.

# Diversity, Domestication and the Death of the Dinosaurs

Having discovered that seeds were an effective way to achieve increased survival of the species, it was only a matter of time before some plants would modify this new method of reproduction. No one knows exactly when it happened, but somewhere around 200 million years ago certain gymnosperms adapted their reproductive methods.

The development of an ovary was land plants' most important adaptation. Within the mother plant's protective ovary, a seed was equipped with as much as it needed to improve its survival rate – the embryo itself, a nutritious endosperm to feed the embryo and a tough outer coat. When the mother plant was ready and her seeds were fully ripe, she could disperse her seeds. After she had released her seeds she could be sure that only under certain conditions, would the seedling break through the outer shell. Just like Goldilocks, it could wait until it was neither too hot nor too cold, neither too wet nor too dry.

The angiosperms – vessel seeds – have

flowers in which to locate the ovaries and all the reproductive parts. The development of a flower and the protection of a seed within an ovary was so important that the angiosperms persist today as the top dogs of the plant world, making up around 90 per cent of the plant kingdom.

Angiosperms were able to spread rapidly through the land once dominated by gymnosperms. Flushed with success, they began to experiment with the shape, colour and structure of their flowers. Sometimes, when pollen was carried from one flower to an entirely different flower, a newer, stronger species was born, one which had both its parents' strengths. This meant they could bypass the slow process of evolution and adapt and diversify with far greater ease. The flowering plants were able to adapt to almost every environment and occupy just about every space from the high forest canopies right down to the ground and everywhere else in between, and all of this in double quick time.

Flowering plants could not avoid being eaten by herbivores, so they turned that fact to their advantage. They produced seeds within edible ovaries. Protected by their hard outer shells, the seeds passed through the guts of animals and were deposited elsewhere. The animals served their plant masters well and dutifully carried angiosperms farther and farther afield.

Angiosperms also found a new way to use animals to carry their pollen from one plant to another. They developed ever-more complex flowers capable of attracting insects to carry the pollen on their behalf.

As the flowering plants discovered different ways to persuade animals to do their bidding, so they had to diversify further by developing new flowers and fruits to suit the land animals. Pangaea – the mass super-continent – had begun to split around the same time as the first flowering plants appeared and this meant that some of the flowering plants found themselves marooned on new continents. Separated by oceans, they were never to meet their relatives again. The same was true of the animals and so flowering plants and animals in each of the new land masses evolved differently and more diversity was achieved.

From dry deserts to vast swamplands, plants were now able to spread just about everywhere. Those dinosaurs that had the good sense to adapt their diets to include the angiosperms, were able to piggyback upon their success and what better place for a young seedling to find itself than in a large deposit of nutritious dinosaur excrement? Land plants and animals lived in relatively peaceful co-existence. However, life on land was soon to experience another mass extinction event that would change everything.

Around 65 million years ago, there was probably at least one asteroid impact and possibly several volcanic eruptions filling the atmosphere with ash. The asteroid's impact added to the problem by generating vast amounts of dust. Anything dependent on the sun was doomed as the sun was locked behind a dark cloud of dense dust and ash. Large animals, needing to eat large amounts of land plants, could no longer survive as their food source began to suffer from the effects of a weakened sun.

It was the extinction event that put paid to the dinosaurs but, though many individual species of plants were lost, the land plants as a group recovered and the angiosperms, with their remarkable ability to diversify, eventually emerged as the confirmed rulers of the plant world. Some of the dinosaurs' descendants evolved into birds and angiosperms were quick to exploit birds, as well as insects, to do their bidding.

Around 11,000 years ago, plants, particularly the angiosperms, found their most powerful ally yet. The 'domestication' of plants by humans was just as useful to plants as it was to humans. In return for a relatively small sacrifice by allowing themselves to be eaten at a certain stage of their lives, plants useful to humans could be assured the survival of their own species. Their progeny was carried all over the world wherever humans went which, ultimately, was everywhere there was fertile land. The most useful plants were protected from the competition and provided with all the nutrition and water they needed.

On the whole, the relationship between plants and humans is so beneficial, it might be argued it was plants that domesticated humans rather than the other way around. Having survived so many turbulent events in Earth's history, plants could survive perfectly well without humans. Humans, on the other hand, would find it impossible to survive without plants.

We are so heavily reliant on plants that life without them is unthinkable. We need them to sustain the very air we breathe, the food we eat and we even have the early plants to thank for the fuel we now use to get us around and to heat our homes. We humans may consider ourselves to be the most powerful beings on Earth, but our lives are incredibly fragile, and though we may now have the tools to weed plants out and the chemicals to kill them, we are but one species in the face of the many hundreds of thousands of species of plants. Some say we are already in the middle of another extinction event. If that's true, it's a pretty fair assumption that human beings will go but plants will simply adapt and survive.

# The domestication of wheat

For those of us who don't have an allergy to wheat, it would be rather difficult to contemplate life without it. Bread, biscuits, breakfast cereal, pasta, pizza, ketchups and beer are just a few of the things we'd have to do without. By sheer volume of production, wheat is one of the most popular food crops in the world. But the wheat we know today doesn't behave anything like the original wheat plant intended. Today's wheat is the result of domestication by humans so that, in the wild, it wouldn't have much of a chance of survival. Just as we need it, domesticated wheat needs us.

Domestication isn't the same as cross cultivation. Rather, it is the selection of certain characteristics that are desirable and the rejection of those that are not. Over time, you get a species so closely linked to humans, it would not survive without us. Domestication creates a situation that is usually mutually beneficial to both parties.

For example, take my dog, Garry. Garry is a border collie with a touch of springer spaniel (one-eighth to be precise). The spaniel part of him was deliberately bred in by the breeder to reduce the natural tendency for border collies to nip at sheep, thus making him more suitable as a pet rather than a working dog. That spaniel bit is the result of cross cultivation – a deliberate attempt to breed in (or out) a particular tendency. The rest of Garry is the result of domestication where traits that would serve him well in the wild have been de-selected over many years to produce an animal able to live amiably with humans. Reduced aggression, lowered predator instinct, together with a total lack of interest in challenging for pack dominance are all traits that make him an ideal pet. However, without aggression and a high predator instinct, Garry, like most of our domestic pets, would have little chance of surviving in the wild.

The domestication of plants by humans also brought about the domestication of humans by plants. In the case of wheat, our hunter-gatherer ancestors had already discovered the benefits of eating the seeds of the wheat plant. They found out they could carry this food around with them without it spoiling like other fruits, berries and leaves. By saving some of the seed and planting it, the hunter-gatherers could finally settle down in one place. They could control where their food source was located without having to waste time and energy looking for it.

Wheat seeds were tricky to collect though. The mother plant had to try to disperse her seeds as widely as possible so the new plants would not compete with one another for light, water and nutrients. She did this by attaching her seeds to an axis (rachis), so delicate when ripe that it shattered in the wind, throwing her seeds far and wide. Once the seeds were detached from the parent plant, they were more difficult for humans to find and many were spoiled by being trampled upon.

Some wheat plants developed a mutation that would have been stamped out in the wild after one or two generations, but was very useful to humans. The rachis of the mutated plant was stronger and the ripe seeds remained on the plant for easy harvesting. The early humans favoured these plants and their seeds were stored and replanted until, over time, all the wheat plants grown were mutants. Without the ability to easily disperse seeds, they were not much use in the wild and almost entirely dependent on humans for their existence.

Wheat and other plants like it changed the development of the human race, allowing us to control our food sources and eventually turn basic settlements into towns and cities. Nowadays, wheat occupies a proportion of the Earth's surface roughly equal to nine times the size of the UK, which makes it pretty successful in the plant world. Not bad for a mutant that should never have survived in the wild.

## When is a flower not a flower?

When it's a cauliflower, of course!

Domestication is beneficial to wheat. A wheat plant gets to live out its full life cycle, to flower, set seed and to fulfil its destiny by producing its next generation. Being an annual plant it can, in effect, die happy. The same cannot be said for the unfortunate cauliflower or indeed for any of its many relatives.

The cauliflower is the result of domestication of the wild mustard plant, *Brassica oleracea,* originally found growing in coastal regions of France and the UK. After its domestication, this plant was interfered with in any number of ways to produce all kinds of vegetables – cabbages, cauliflower, Brussels sprouts, broccoli, kale and kohlrabi to name but a few. For all of them, domestication and breeding have resulted in plants that are usually prevented from reaching their true potential and producing flowers, seeds and the next generation.

In the case of cauliflower, it is thought cauliflowers are the result of a mutation causing the cells which should produce a flower to constantly generate replicas. This means you get a dense cluster of under-developed flowers and is what gives cauliflowers that fractal effect where each smaller piece you break off looks like an exact replica of the bigger piece and so on. As a result, the flower is not a true flower at all but a cluster of aborted flowers that will only produce true flower stalks in the second year (presumably after it has tired of producing pointless non-flowers). Sadly, cauliflowers never get to realise their dreams of producing a true flower because they are usually culled when they're less than a year old.

Domestication has been quite hard on the poor wild mustard plant and the result of all that domestication followed by artificial selection over centuries has resulted in a group of vegetables requiring more than the usual amount of human interference to grow. The relative difficulty of growing plants from the brassicas group is a reminder to us that, just as with animals, if we domesticate plants, it's really no wonder they come to rely on us for their survival.

Because kale has been allowed to retain the most natural growth pattern, it is the easiest of all the brassicas to care for. The rest of the group is attacked by all manner of pests and diseases – caterpillars, aphids, thrips, birds, bacterial black rot and a wide range of fungal diseases of which club root disease is probably the most common.

If you want to grow brassicas, you need to take on board that these mutant plants have virtually no survival skills of their own. Nature might even consider them an abomination and would probably wipe them out if it could. As a result, they will require your protection at every stage of their short lives. Lime the soil before you plant them to replicate the chalky soil on which their wild relatives grow. This helps to guard against club root disease. Keep your brassicas netted at all times with good insect netting to prevent cabbage butterflies from laying their eggs and stop pigeons from pecking at the young plants and stripping the leaves of older plants. Never grow brassicas in the same place for two years running to prevent the build-up of diseases and pests.

For cauliflowers, it is important you keep them growing steadily because, if their growth is checked at any time, they will produce stunted and distorted heads. This means making sure they are watered regularly and protected from extremes of temperature. During really cold weather, they will need frost protection and during really hot weather, they should be shaded from the sun. They are very greedy plants and require a rich, fertile soil and further application of high nitrogen fertiliser once the plants are growing well. To keep the head nice and white, you should bend the outer leaves over the curd and tie them so you get a clean head without any sunburn. You need to cut them at just the right time too – when the heads are firm but before the florets have begun to separate.

# Part Two: Sex on the Brain

# Plants on the Pull

All plants are capable of sexual reproduction. This ensures survival of the fittest and allows them to successfully evolve and adapt. These days this seems like a pretty obvious statement, but the notion that plants engaged in sexual activity was strongly denied by some religious folk, right up until the 19<sup>th</sup> Century. The very idea that plants had sex! Imagine the prudish spinster lady of the parish bending over to smell the scent of a lily only to be reminded she was rubbing noses with both male and female sex organs? The shame of it!

Of course we now know sexual reproduction in plants takes place in much the same way as it does in other living species, including humans. The male part of the plant interacts with the female part and you get a progeny that is a genetic mix of both parents. Unlike us, one plant can't simply approach another plant and proposition it. Unable to seek out a sexual partner for themselves, plants rely on other living things, or on the wind or water, to carry the male pollen to a receptive stigma (the female part of a plant). Though it would be nice to think flowers are on this earth to give us pleasure, this is just a happy side-effect of one of the ways plants connect with the opposite sex. Most brightly coloured flowering plants use go-betweens such as bees, bats, birds, flies or butterflies to carry the pollen to the stigma. It is these animal pollinators the flowers are designed to attract.

So, how does a flowering plant prepare to go out on a date? Well, it does it more or less the same as we do. It changes its appearance, wears attractive clothing, puts on its makeup and some even apply a little scent. By the time it's finished, it looks a lot different to the way it used to look. Step one in the plant dating game: make yourself as attractive as possible. All made up and dressed in its finery, the plant is sending out a clear signal. I'm here and I'm sexy, so come and get me!

As a plant approaches sexual maturity, it spends its days collecting and storing energy, waiting until the time is right to enter the dating marketplace. The flowering season is such an important event in the plant's calendar that it will spend a great deal of time making sure it's ready. A weak plant will only be able to produce weak flowers, so it tries to gather as much strength as possible.

It would be pointless to waste energy producing a flower if there were no potential sexual partners to mate with. Each species of plant has its own set of boxes to tick before it will begin the arduous task of producing a flower. For example, spring and autumn flowering woodland plants must take advantage of the time when the leaves are off the trees allowing light to reach the plants below. They must also be sure any pollinating insects have woken from winter hibernation so they carefully record both the length of the nights and the temperature. If they flower too early, it won't be warm enough for any pollinators to be around and, if they leave it too late, the leaves will be on the trees and there won't be enough light to give them the energy they need to produce a flower. In autumn, it's the other way around – too early and there won't be enough light, too late and it'll be too cold. This carefully timed attitude to dating explains why we can't persuade our plants to flower exactly when we want them to unless we artificially alter their environment. By adjusting the light levels, the temperature and other factors, we can fool plants into thinking it's time to flower, making it possible to display snowdrops at Chelsea Flower Show at the end of May.

Under normal circumstances, only when all the factors such as correct temperature, suitable night length, intensity of light and amount of rainfall are present is the time right for the plant to switch on the flower-making procedures. It will then set aside certain cells to develop into flowers forming first the protective green sepals, then the coloured petals and finally the sexual organs inside. Once the flower is built, it will open to signal it is now sexually active.

As you might imagine, all that dressing up is exhausting and some flowering plants really can't be bothered. These plants use the wind to disperse their pollen, throwing out huge amounts of microscopic pollen in the hope that some of it will be lucky enough to land on a receptive stigma of the same species of plant. This is a fairly chaotic method of sexual reproduction and not as efficient as pollination by others so the plant is required to produce more pollen to achieve success. These are the old-fashioned members of the flowering plant world, preferring to stick to methods used millions of years ago rather than rely on more modern flower production.

The pollen of wind pollinated plants is completely indiscriminate about where it lands and while it may not be a particularly nice thought that, at certain times of year, we have plant sperm up our noses, this is invariably where some of this pollen ends up. Most trees and grasses are wind pollinated and people with an allergy to the pollen of either or both will curse these plants.

You can nearly always tell just by looking at the flower of a plant, whether it's animal or wind pollinated. Wind pollinated plants don't need to attract passing animals so why bother putting all that energy into forming a complex and colourful flower? Likewise, there is just no point whatsoever to wasting energy on producing perfume. Their flowers are relatively drab – often the same colour as the leaves. Their pollen is located on fine stalks (filaments) that move easily in the wind sending the pollen grains flying through the air with their fingers crossed they will land on the right receptive female. Have a look at the flowers on any of the grassy plants and you'll see what I mean.

If wind pollination is such a chaotic method of finding a sexual partner, it makes you wonder why some flowering plants have stubbornly refused to ditch it. Well, while making yourself attractive to a certain group allows you to be more selective, there is also an advantage to remaining drab and dreary. For something unable to run away from predators, producing a showy, scented flower may get you noticed by the wrong type of insect or animal.

Whether your plant uses wind or animal pollination, the production of a flower represents the beginning of its lifelong dream to achieve sexual reproduction and ensure the survival of its species into the next generation.

# Giving wind pollinated plants a hand

In the vegetable garden, there aren't that many wind pollinated plants of interest to us in terms of developing or encouraging flowers. Beets, chard and spinach are all wind pollinated but, since we are not interested in the seeds or fruits of these plants, successful sexual reproduction is of no concern to us unless, of course, we want to save the seeds to sow the following year. With sweet corn, however, we need a transfer of pollen to form the kernels. In other words, the plant must achieve sexual reproduction so it can produce seeds for us to eat. Because wind pollination is so hit-and-miss, we can increase the chances of success by either pollinating by hand or by growing the plants in blocks, rather than rows.

Sweet corn is monoecious – it produces a separate male flower and different female flowers on the same plant. For sexual reproduction to be successful, the pollen from the male flower at the top must land on the female flowers underneath. Once that happens, the female flower can get on with the task of producing the seeds. Plant your sweet corn, not in rows as you might other vegetables, but in blocks, laid out in a four plants by four plants square. This will increase the likelihood of the pollen from the tassels (the male flowers at the top of the stems) landing on the silks (the developing female ears farther down).

If you don't want to trust your sweet corn-growing success to luck, you can force a match by stroking first the male tassels and then the female silks. Having done this myself, I can tell you it's actually quite soothing to stroke a sweet corn. Who needs a cat?

# Deadhead to prolong the flowering period

Our flowering plant has spent all that time and energy making itself more attractive. It's an arduous task and once it is successful at finding a sexual partner, it doesn't need all that make-up any more. As soon as it's been fertilised, it's going to require all its strength to make it through pregnancy. You'll know when a flower has become fertilised because the pretty petals you love so much will shrivel up and drop off. As soon as you see this happening and, before too much energy is directed towards seed production, you need to remove the spent flower.

Some of our most popular plants are bedding plants. Every year we spend a fortune on trays of these beauties to provide seasonal colour for our gardens. Bedding plants are chosen for their bright colours, relatively predictable height and spread, and prolific flowering habit. There's a very good reason these plants produce such abundant and eye-catching flowers. Almost all our bedding plants are short-lived. Most will not make it through winter so they're in a real rush to find a sexual partner. As a result, they put all their energies into producing as many flowers as they possibly can before they die. Once they've been pollinated, they will slow down or cease flower production to concentrate on the all-important seeds.

Deadhead bedding plants as often as you can by pulling off the wilted flower heads with your fingers or use secateurs if the stems are too thick. When you do this, and providing there's still enough time before winter sets in, the plant must reach for the make-up bag all over again and you'll get more flowers for longer. You can deadhead any flowering plant and, in most cases, you will be able to prolong their flowering period by several weeks. It doesn't really matter where on the flower stem you make the cut so long as you remove the flower. For very fine plants with lots of small flowers, you can just take some shears to the whole plant and, although it will hate you for ruining its efforts, it will resentfully reward you with a host of new flowers.

# Choosing the Right Matchmaker

As soon as a plant has produced a flower, it's ready to engage in sexual activity. Since both it and any possible partners are stuck in one place, it's not a question of waiting for a knock on the door from a potential suitor so many plants must use a go-between to give love a helping hand. The most efficient and successful way to achieve pollination is to use animals as matchmakers. Over the years plants have been very clever at adapting their flowers to attract whatever animals have evolved alongside them.

Plants attract their pollinating partners not only by looking a certain way, but also by offering a reward. What do you do when you see someone you like in a nightclub? You offer to buy them a drink, of course! For most plants, the drink in question is nectar and a whole host of animals are unable to resist it.

To find a matchmaker, plants must advertise their presence - a risky business for something unable to run away from predators. If plants can narrow the field to attract only certain types of animals or insects, they can reduce the risk of being found by anything that might harm them.

Choosing certain matchmakers over others has another benefit. A pollinator must typically carry the pollen from one plant to another plant of the same species in order for fertilisation to take place. By attracting only certain types of go-betweens, the entire species can reduce the risk of wasting pollen on plants that aren't a proper match.

Plants come in all shapes and sizes. A cursory look at a group of plants in a border when they're not in flower might lead us to believe they all look the same. On closer inspection, however, they are very different in their growth habit, leaf shape, leaf arrangement and colour. However, when they flower we can really see, at a glance, how very diverse they are and much of that diversity is to do with the ingenious way they use flowers to attract pollinators.

Bees seem to have a preference for blues over reds. The general consensus is that they can only see colour on the blue end of the spectrum. Red, for example, will appear as black. When they see yellow, it appears as blue, so yellow and blue flowers have an advantage. Bees also have an excellent sense of smell. Plants can emit a delicate scent that will be sure to attract a bee, but may not be picked up by other animals. Bees have short tongues so it wouldn't do for a plant that wants to use bees to hide its nectar deep in the flower. A scented flower where the pollen is readily visible will, in all probability, be trying to attract bees. Some plants create a landing platform that is highly visible to the bee. Just as helipads have a giant H painted on them to guide the pilot to the landing pad, so some plants will create a circle of contrasting colour around the centre of the flower, rather like a bull's-eye, to guide the bee. Others, like foxgloves, have lines of spots at the entrance to the flower like lights on a runway, guiding the bee straight to the nectar source.

Butterflies and moths have very long proboscises – appendages they use like a straw to suck up food. Plants wanting to attract butterflies hide their nectar deeper so that only butterflies can reach it. Plants with long tubular flowers are looking for butterflies, but hummingbirds can reach nectar deep in the flower with their long beaks. How then do plants attract butterflies whilst avoiding hummingbirds? Birds have very little sense of smell so plants seeking to attract exclusively butterflies may also use scent to improve their chances.

At dusk, plants hoping to attract moths and bats open their flowers. Most flower-visiting bats inhabit tropical and desert climates and, when you are able to attract bats, plants get the added bonus of a little pest control thrown in since the bats also consume unfriendly insects. Another plant pollinator with a duel role is the hoverfly. Many species of hoverfly feed on nectar, but their larvae feed on aphids and other sap-sucking pests. Any plant able to attract these insects not only gets pollinated, but, while visiting the plant to feed, the hoverfly might lay some eggs for future pest control. It certainly pays to date a hoverfly!

Flowers you wouldn't want to put in a vase on your dining table belong to those plants that prefer to attract certain types of flies for pollination. These plants have flowers that not only look like rotting flesh, but smell like it. They attract flies that feed on carrion. Not the kind of plants we would want to grow or eat, but why should plants care what we want? Smelling putrid and looking unappetising has obvious benefits to a plant if it wants to be pollinated by a fly. Since the flower gives no reward whatsoever to the fly, this is very much a one-sided relationship.

Just like some humans, plants can be very good liars when it comes to sex. Some plants will even go so far as to produce flowers that look and feel like a female insect. Along comes a passing male, thinks he's onto a good thing and attempts to have sex with the flower, at the same time picking up some pollen for his troubles. But how do plants figure out these things in the first place? They can't see a female insect in order to impersonate it. How do they know they're supposed to have yellow petals or blue petals or to emit a certain smell?

Well, it's certainly not a deliberate decision by the plants, they're not that smart. While they may not be Einsteins, they are most definitely scientists and they are scientists with an amazing penchant for experimentation. In the case of the insect impersonator, some plant happened to produce a new hairstyle and, by happy accident, it suddenly found itself very attractive to a certain type of male insect. This union proved so successful, the plant in question was able to out-compete others by producing more offspring. The offspring inherited the new hairstyle until eventually the fashion became a must-have.

For all sorts of reasons, it's a great advantage for a plant species to be able to choose its matchmaker and we can use this to our advantage. If we know which plants attract which types of pollinators, we can encourage our favourites to enter our gardens.

## Sexual promiscuity can lead to STDs!

Humans know that unprotected sex with indiscriminate sexual partners can have disastrous consequences. So how come plants get away with all that fun? Well, actually, they *don't* always get away with it. Being so promiscuous can lead to consequences for plants too.

Plants can and do get sexually transmitted diseases and it's all down to bad luck. One such

STD is the brilliantly named 'Anther Smut Disease' which infects Moss Campion. Insects carry pollen from infected plants to unsuspecting female parts of other plants, thus transmitting the disease. Anther Smut is a fungal disease affecting the male organs (anthers) of the flower, turning them black and shrivelled. Like many sexually transmitted diseases, it causes sterility in plants and makes them look pretty unattractive as well. Again, like other STDs, the physical symptoms don't appear on the plant until the following year, so plants can be carriers without any obvious symptoms. By the time the physical signs appear, it is too late.

## Planting a bee and butterfly café

Einstein is reputed to have said: "If the bee disappeared off the surface of the globe, then man would only have four years of life left."

Since we know how adaptable plants can be, perhaps this is a bit of an exaggeration, but there's no doubt that bees are vitally important to life as we enjoy it today. The current global decline in bee populations is a real worry for commercial growers and, if we have to resort to hand pollination in place of the bees, our food will become far more expensive.

As well as being extremely pretty to look at, butterflies are an important food source for birds so every gardener should consider it his or her duty to open a bee and butterfly café somewhere in their garden. Here's a list, though not exhaustive by any means, of what you could have on your café's menu:

| **For bees** | **For butterflies** |
|---|---|
| *Aquilegia* | Bluebell *(Hyacinthoides)* |
| *Aster* | Broom *(Cytisus)* |
| *Aubrieta* | *Buddleia* |
| Bergamot *(Monarda)* | *Coreopsis* |
| Borage *(Borago)* | *Cosmos* |
| Bugle *(Ajuga)* | Daylily *(Hemerocalis)* |
| Clover *(Trifolium)* | *Echinacea* |
| Cornflower *(Centaurea)* | Fennel *(Foeniculum)* |
| *Cotoneaster* | Hollyhock *(Alcea)* |
| Foxglove *(Digitalis)* | Lilac *(Syrimga)* |
| Honeysuckle (*Lonicera*) | Marigold *(Calendula)* |
| *Lavandula* | Nasturtium *(Tropaeolum)* |
| Lemon Balm *(Melissa)* | Nettle *(Urtica)* |
| Marjoram *(Origanum)* | *Penstemon* |
| *Pulmonaria* | Snapdragon*(Antirrhinum)* |
| *Rhododendron* | Sweet *Alyssum* |
| *Rosemarinus* | Sweet pea (*Lathyrus*) |
| *Scabiosa* | Thistle *(Onoropordum)* |
| *Thymus* | *Verbena* |
| *Wisteria* | Yarrow *(Achillea)* |

# Hand pollination for growing vegetables and fruit

If we are growing fruit and certain vegetables, we want the plants to achieve pollination so the female part of the plant can get on with the business of developing seeds and a fruit. Insect pollination is very efficient and plants are so good at it, they don't really need our help. If you want to feel part of things, you can give love a helping hand by pollinating the flowers of your fruit and vegetables manually. To do this, use a soft brush or feather to brush the pollen from the anthers of one flower and deliver it to the stigma of another. Simply brush the centre of the flower on one plant and then brush the centre of the flower on a different plant of the same type – job done!

You can do this with courgettes, squashes, beans, peas and any kind of fruit. Even fruits and vegetables that are self-pollinating will benefit from a little help. Tomatoes, for example, don't need to find a sexual partner for seed production and you only need one plant to produce fruit. However, they still require some movement to dislodge the pollen so it drops onto the stigmas. Outdoors, either the wind or insects will do that for them but, in a greenhouse where there is lack of air movement and low insect numbers, a little brush with a feather won't do them any harm at all.

Hand pollination is particularly important for plants grown in greenhouses and, outdoors, during summers when the weather is warm and wet. Bees and other flying insects don't like rain so, if there are too many rainy days, you might have to do the job for them. During a wet summer, if you notice some of the fruit on the likes of courgettes and squashes withers away without fully developing, this could be a sign that pollination hasn't occurred and the plant has given up on that particular fruit. If you get to work with a brush or a feather, you should be able to save the rest. You'll find the male flowers on a long, thin stem. The female flowers have a shorter stem, sometimes with a small fruit developing behind the flower. Brush the male first and then the female. Again, where possible, you should brush the male from one plant and transfer the pollen to the female on a different plant.

# Avoiding Incest

There's a very practical reason why it's never a good idea to have sex with a close relative. When an offspring is produced from genetically similar parents, you run the risk of passing on less desirable physical traits that might otherwise be cancelled out. Undesirable genes are nearly always recessive – in other words, they can be cancelled out or diluted by stronger genes. If both parents have the same recessive gene because they are closely related, this means the recessive gene is able to continue unchecked. The same rule applies to plants. Plants can and do self-pollinate and this will lead to self-fertilisation. *Bougainvillea* plants wait for just one day for a visiting insect or hummingbird to pollinate the flowers. If this doesn't happen, the following day, the floral tube twists so it can pollinate itself. Not the best result for natural selection, but a result for the *Bougainvillea* all the same.

For the most part, inbreeding in plants is undesirable. It is far better for plants to cross-pollinate. In other words, a plant prefers to breed with another plant of the same species, rather than with itself and, in general, plants that are manually forced to self-fertilise produce weaker offspring.

So, how does a plant avoid breeding with itself? Most flowering plants are hermaphrodite – they produce flowers with male and female parts contained in the same flower. The best way to see a hermaphrodite flower is to look into a daylily or a *Fuchsia* flower. In these flowers you can clearly see the cluster of male parts surrounding a single female. On the stalks of the male parts is the dust-like pollen. The female part has a little blob on top of the stalk (the stigma), sticky enough to capture the pollen when it lands on it.

When you watch a bee climbing over the male and female parts of a hermaphrodite flower, it seems impossible for the plant to avoid mating with itself. Hermaphrodites avoid self-pollination by ensuring the female part is ready at a different time to the male parts. That way, if the male pollen comes into contact with the female stigma on the same plant, the female won't be receptive to fertilisation. Only when the pollen lands on a different plant of the same species whose stigma is receptive, can fertilisation take place. Another way plants avoid their pollen coming into contact with their own stigma is to make the female stigma longer than the male stamens so there's less chance of an insect climbing over both.

Monoecious flowering plants are those that produce separate male and female flowers on the same plant. This ensures the pollen is far enough away from the female parts to reduce the risk of self-pollination. Sweet corn is a good example of a monoecious plant. Just to be sure there can be no mistakes, the male and female flowers on each individual plant are ready at different times.

Dioecious plants produce entirely different plants – a male plant and a female plant. Dioecious means 'two houses'. In the case of dioecious plants, think of a fraternity and a sorority – the males in one house and the females in another. To achieve fertilisation, and ultimately seed and fruit production, the male and female plants of dioecious species need to be close enough so successful pollination can occur. In most cases, if you want fruits to appear on your dioecious plant, you need to make sure you have a male and female plant in close proximity.

Just in case accidental self-pollination does occur, many flowering plants have the ability to reject their own pollen by performing self-abortion to ensure a possibly weak offspring is rejected.

## Weeding out the males

Arguably the most famous and controversial dioecious plant is cannabis. The psychoactive and medicinal substances in a cannabis plant are contained in sticky, resinous glands (trichomes) and these are more abundant on female flowers. When the female's flowers are pollinated, she diverts her energies into producing seeds. If the female plant is not pollinated, she continues to mature and produces a larger flower and more trichomes. Therefore, cannabis growers must identify the males and remove these plants to avoid fertilisation of the females.

The longer she has to wait for pollination, the more trichomes a female cannabis plant will produce. No one knows the definitive reason for this, but it could be that she is making herself more 'sticky'. Since these plants are wind-pollinated, the female flower needs to be sticky enough to catch any pollen that might be floating by. The longer she has to wait for pollen to arrive, the stickier she has to get to increase her chances. It may be that she is producing trichomes to use as a kind of sun block, protecting the precious flower from harmful UV rays. This theory would also make sense since the longer the flower is forced to wait around in the sun, the more protection the female might need.

Whatever the reason for the production of trichomes it has made female cannabis plants especially popular with some humans and we've been cultivating this plant since at least 2200BCE. And as for the males? While many of them are put to death, some are spared to facilitate the production of seed.

# Why doesn't my holly have berries?

If we choose to grow a holly in our garden we usually do so because we expect to see red berries in winter and it can be disappointing when these elusive berries don't appear. Holly trees are dioecious and, under normal circumstances, a female plant needs to be within around 30 metres of a male in order for it to achieve pollination and therefore subsequently produce a seed surrounded by a fruit (the berry). If you want to be really pedantic holly berries are not true berries, but drupes, since the flesh only surrounds one seed or stone. However, we'll stick with calling them berries since that's how most of us refer to them.

The best way to be sure you get berries on your holly is to plant at least one male close to your females. Just as it is with humans, one male can pollinate and ultimately fertilise several female plants.

It's not always easy to tell the difference between a male and a female. If it has berries, it's a female. The flower in a male has stamens with no central stigma and style. The flower in a female has a central stigma and style with no surrounding stamens but you don't often buy hollies when they are in flower or fruit. If you buy named cultivars, you can usually tell by the name – 'China Boy', for example, is male and 'China Girl' is female. Just to confuse you, one of the most popular varieties of holly –'Golden King' - is female and 'Golden Queen' is male! If in doubt, visit a good plant nursery and ask their advice – they'll keep you right.

Alternatively, you could buy a parthenogenetic holly – parthenogenetic meaning 'virgin creation'. A few cultivars of holly are available that do not require a male to produce fruit but where's the fun in that?

Hollies don't generally reach sexual maturity until they're around three to five years old, so if you buy a really young plant, you might have to be patient for a while until the berries make their first appearance.

## Successful apples

If you want to grow apples, you need to remember that plants do everything they can to prevent self-fertilisation. Apple trees are hermaphroditic, but are generally known to be self-incompatible, meaning that most varieties will stubbornly avoid self-pollination. If you don't want to leave it to chance that another apple tree might be growing close by, you should buy at least two different varieties of apple tree. Just because you have two different varieties doesn't necessarily mean they will be compatible since some varieties will breed together better than others. It's all to do with flowering time. You need to make sure your varieties are properly paired up so that they flower simultaneously. To guide you, apple cultivars are divided into seven groups depending on their flowering time. You need to pick two trees from the same group for the best results.

When you shop for apple trees at a good nursery, the label will list the varieties that should be grown alongside for the best results. Alternatively, you could buy specially bred self-fertile varieties. These apple trees will fertilise themselves and will at least produce fruit for you to eat. Only, don't plant the seeds because you won't be doing natural selection any favours.

It's not only apple trees that avoid self-pollination. Pears, sweet cherries and plums should all be grown alongside compatible partners for the best results.

# Messing around with Nature

At any given time, we have a number of pollinating insects flying around with pollen stuck to them. Add to this, all the pollen floating on the wind just looking for a place to settle. With all this pollen kicking about, how is it possible for plants to breed only within their respective species? Why does a rose not breed with a petunia? In the case of roses and petunias, roses belong to the genus, *Rosa* and petunias belong to the genus *Petunia*. With roses and petunias, the answer is relatively simple. They can no more breed with ease than humans can breed with horses. It's one of those laws of Nature. Humans belong to the genus *Homo* and horses belong to the genus *Equus* and, unless some mad scientist manages to meddle successfully with Nature's laws, centaurs will always remain the stuff of myths.

*Homo sapiens* are the only species left within their genus. This makes us relatively rare amongst other living things. For example, within the genus *Rosa,* there are more species than I care to mention. Because they diversified in such a fragmented way, separated from their close relatives by their turbulent pasts, it is possible to find plants that are different species, yet have similar characteristics.

The genus, the first part of a scientific name, tells us that all species within that genus possess quite similar physical and genetic characteristics. The species, the second part of the name, tells us that the broad characteristics they share are identical. Living things that belong to the same genus and species, for example, Homo *sapiens,* are capable of breeding with one another with relative ease.

If plants could see, all humans – blond, brunette, white or black - would look the same. Regardless of the colour of our hair or skin, we all walk upright, have forward facing eyes, one heart, two lungs, one nose, one mouth and so on. On the other hand, one species of geranium may look the same as another to you and me but they are very different. Like the genus *Rosa*, there are many species within the genus *Geranium* and each species is more different than all humans are to each other.

Despite their differences, it is possible for separate species within the same genus to breed and produce successful progeny. A mule is the result of breeding between a donkey and a horse – made possible because they are both at least from the same genus. Ligers – crossbreeds between lions and tigers - are both from the genus *Panthera* so their cross breeding, though very rare, is possible. However, there's a reason why you don't see loads of mules and ligers in the wild. Quite simply, most species prefer not to mate with others not of their own kind. Even when it does happen, Nature often puts a stop to it by making the resultant hybrid sterile. If you want to make another mule you can't breed two mules, you have to go back and breed a horse and a donkey again. Ligers, too, are sterile.

Some plants within the same genus, but of a different species, may hybridise of their own accord and plants are more capable of hybridising than other living things. Take two species of *Geranium* whose natural habitat may be hundreds of miles apart. Put them together in a garden and they may well breed with one another. This allows us to experiment with breeding plants and produce some interesting results. We can produce plants with bigger blooms, better yields and increased resistance to disease by forcing cross-pollination between species of plants that, if left to their own devices, would never normally get together. But, as much as we'd like to think so, we're not God and Nature makes sure she brings us down to earth when the plants we produce through hybridisation revert to one or other of the two species they were bred from.

Take, for example, two plants of the same genus that are hybridised – one is a strong plant with good resistance to disease, but slightly misshapen fruit, the other is not quite so strong but produces a more uniform fruit. Successful hybridisation may produce a plant that inherits good disease resistance from one parent and uniform fruit from the other. However, there is a strong possibility that, if we collect the seed from the hybridised plant, the strongest gene will dominate once again and we will get a return to better disease resistance with fruit that doesn't look quite so good on the shelf.

So while it is possible to mess around with Nature, and very often we do, the laws that maintain survival of the fittest are often just too great a force for us to overcome permanently.

## Frankenstein is born

In 1888, on the Leyland Estate in Powys, Wales a Monterey cypress from California met a Nootka cypress from Alaska. The resultant offspring was a monster – a fast-growing, thick conifer capable of growing around a metre a year and able to withstand poor soil conditions, wind and salt-laden air. Leyland cypress had inherited the speed of its father (Monterey) and the hardiness of its mother (Nootka).

We love this plant. As a hedging plant it establishes quickly and we can grow it in places where other conifers would fail. However, we forget that it is, in fact, a monster. It grows so fast it blocks out light and can even overshadow whole houses. The plant caused so much controversy it forced the introduction of a law, in England and Wales, nicknamed the 'Leylandii Law', allowing local authorities to legislate for people to reduce the height of hedges that block out their neighbours' light. In 2001, a man was shot dead by his neighbour. The two were alleged to have been in dispute over a *leylandii* hedge. Ironically, they lived in Powys where the monster *leylandii* was born.

Luckily, Leyland cypress is sterile. All the Leyland cypress you buy are propagated from cuttings so we can rest easy knowing this plant will never be able to colonise our space. Imagine what would happen if the monster *leylandii* started popping up all over the place and was able to take over spare ground and empty spaces with no one to check its growth. So even if we don't know what we're doing, Nature nearly always does.

# Who do you think you are? Michael Schumacher?

Peppermint is the result of crossbreeding *Mentha aquatica* (watermint) and *Mentha spicata* (spearmint) yet it refuses to become a fully functioning species in its own right. Time and time again, two peppermint plants that mate produce sterile seed. Peppermint is an F1 hybrid. And no,

F1 doesn't stand for Formula One but for Filial One, meaning the first generation produced by cross-mating two distinctly different parents. If you buy seeds, you'll often come across this term.

F1 hybrid seeds don't usually produce plants whose seeds, in turn, will produce plants identical to the original F1. When this is the case, we say the seeds won't 'come true'. To reproduce the characteristics of the original seeds, you must go back to the original parents and breed them again.

If you buy F1 hybrid seeds, you are usually buying seeds that will produce plants of excellent quality. Often, the seeds have been cross-pollinated by hand under controlled conditions so you'll get a lot less seeds for your money. Don't save the seeds from those plants – you'll probably be disappointed with the results. F1 hybrids should always be grown with seeds bought from suppliers, not with seeds you collect yourself.

If you want to try creating your own hybrid plant, you need to choose two compatible plants. For best results, these should be two species of plants belonging to the same genus. Maybe you want to cross two different coloured geraniums to see what you get.

If the flower is hermaphrodite, before the flower of your chosen 'mother' plant is fully open, you need to carefully separate the petals and snip off the stamens so there is no chance of the plant's own pollen reaching its stigma. Once you have done this, cover the flower with a paper bag to prevent any stray pollen from reaching it. You should also cover the donor pollen plant so there is no contamination of the pollen.

Your best chance of guessing when the female part of the mother plant is receptive is to touch the stigma to see if it feels a little sticky. You can then remove the pollen from the father plant using a brush or feather and lightly dust the stigma with pollen.

Cover up the mother plant again and label it with the names of the two plants you have crossed. After that, it's just a case of waiting to see if the plant produces seed. You then have a further wait for the seed to germinate and even longer to find out how the new cross behaves over a few generations. If you manage to produce an entirely new variety, you get to name it and may even get a royalty every time it's sold.

## Making sense of plant names

All known species are named according to a two word Latinised system known as the binomial system. The hardy Geranium, *Geranium endressii* is similar but not identical to *Geranium himalayense*. In fact, there are over four hundred species within the genus *Geranium*, all of them similar but each of them different enough to merit being named as individual species. *Geranium endresii* and *Geranium himalayense* are true species and, as such, they reserve the right to be known by a two-word name. Their first name being the genus they belong to and their second, being their species.

Plants of the same genus and species can display slight variations from the true species in the same way humans have different hair, skin and eye colour. If this occurs without human interference and, if the seeds breed true, these plants are sub-divided into named varieties. In this case, a third word is added that is usually Latinised and proceeded by the shortened word var. For example *Geranium pratense var. stewartianum* is a coarser-leaved, pink-flowered variation of the cranesbill, *Geranium pratense*. These plants are still part of the same species but with a very slight difference.

Cultivar is short for cultivated variety. Generally, these plants are not naturally self-sustaining like the varieties and often have to be propagated by vegetative propagation. Cultivars can be the result of deliberate hybridisation or an accidental mutation but, in all cases, they are maintained by humans because natural selection wants nothing to do with them. You can tell if your plant is a cultivar if its species name is followed by another, encased in single quotation marks. An example of a cultivar is the white flowered *Geranium pratense* 'Galactic' which is a cranesbill with a variation bred into it.

Cultivars and varieties are generally variations within the same species. However, like ligers and mules, it is possible to cross plants of different species within the same genus (an interspecific cross). We can force this to happen or we can leave it to chance and try to exploit the results when we come across them. The ability of flowering plants in particular to mate fairly readily was, of course how they achieved all that diversity so quickly millions of years ago. When a plant is the result of an interspecific cross, an x is inserted into the name, for example, *Geranium x cantabrigiense*. If the parent plants are of different genus (an intergeneric cross), an x is inserted in front of the Latin name, for example, *x Fatshedera lizei*. Intergeneric hybrids are pretty rare.

You can also get cultivars of crosses and cultivars of varieties. Sellers and plant encyclopaedias like to shorten plant names for convenience, especially if they have been around for a long time and have been crossed many times. *Geranium phaeum var. lividum* 'Joan Baker' might be shortened to *Geranium* 'Joan Baker'. This certainly makes it more convenient to write down and to say, but not so convenient for us if we want to understand the origins of our plants.

I am often asked why we bother with Latin

names for plants. As you have seen, the correct name of a plant can tell you a lot about it. More than that, giving a plant its true name helps it achieve worldwide recognition. By using the Latin name, you can order any plant from any plant nursery in the world and they will know exactly which one you're talking about. In the case of cultivars and varieties, you can be assured you will get the exact shape and colour you want.

# Pregnant Plants

Pollination is the same as the act of sex. Only when the female part of the plant is ready to receive the pollen and when both plants are compatible, does plant sex result in pregnancy. In plant terms, a positive pregnancy result means it has achieved fertilisation. The flower petals fall and the ovaries swell up with one or more embryos, each of which is encased within a seed.

All flowering plants produce seeds contained within an ovary which protects them while they mature. Pregnancy in plants can last from one week to several years, depending on the species. During this time, most of the plant's energy is concentrated on developing the precious seeds. It often continues producing flowers as a back-up plan in case the first batch of seeds fails, but its flower-production will be reduced.

Inside the seeds, embryos develop surrounded by tissue to provide food and give them the energy they need to break out when the time is right. In monocotyledons, or monocots for short, the embryo contains one seed leaf and, in dicotyledons (dicots) it has two seed leaves. In the case of monocots, the embryo will use the packed lunch that surrounds it when it germinates. In the case of most dicots, the packed lunch is eaten before the embryo breaks out of the seed. You can tell if a plant is a monocot or a dicot by looking at the leaves. The leaves of monocots have parallel veins and dicots have net-like veins.

The ovary itself matures along with the seeds it contains. Some may become fleshy like tomatoes and some may harden like nuts. Often these mature ovaries are edible. When you eat a tomato, you are basically eating an ovary.

As the flowering plant reaches the final stages of pregnancy, the coating surrounding the seed hardens to prepare the seed for life outside the ovary. At this stage, the embryo stops growing and becomes inert, so it can remain in a dormant state until the time is right to burst into life. The seed is now ready to leave the protection of its mother's womb. Each seed contains a new life and each seed is an individual, a genetic mix of its parents with its own unique combination of its parents' genes.

Some plants wear their pregnancy better than others. Those plants with edible fruits like tomatoes and peppers look positively glowing with their shiny red, green, yellow and orange bumps. Others hide their pregnancy behind wilted petals and brown seed heads, looking as if they are absolutely exhausted by it all and causing us to think they might be dying. They are not dying, merely pregnant.

## And you thought nine months was tough

I first came across a real-life specimen of the world's largest seed whilst on a visit to the Eden Project in Cornwall. Staring down at the seed of the coco de mer palm tree, which looked rather like a well-rounded bottom sticking out of the soil, I must admit to thinking to myself, "So what?" At between 40 and 50cm it certainly is big, but what's more impressive than the seed itself is the mother plant that produced it. The female coco de mer takes between five and six years to bring its fruit to maturity. So now I'm impressed. That's a hell of a long time to be pregnant! The male coco de mer plant is no less impressive, with a wonderfully phallic flower, between one and two metres long.

So, we have a female plant bearing a seed that looks a lot like a female's buttocks, and a male plant with a magnificently phallic flower. How could there not be an ancient legend about this rare tree? Well, legend has it the male trees uproot themselves and stride over to the females for a night of passionate lovemaking. There are, of course, no witnesses to this event because anyone who does see it, is struck blind or dies! The seed is also thought to have given rise to tales of mermaids. Sailors catching sight of coco de mer seeds floating on the sea could be forgiven for thinking they were looking at a mermaid's bottom.

## Antenatal care for your plants

Giving a plant everything it needs to help it through pregnancy will reward us with better and bigger fruits. By fruits, I don't just mean sweet-tasting fruit like raspberries and strawberries. Pumpkins, courgettes, peppers and tomatoes are fruits and so are pea and bean pods. Once your plant has begun to produce a fruit, you can start administering some TLC to help it through pregnancy.

Plants need leaves to collect energy from the sun and this is especially important during pregnancy. However, some plants don't know when they've got too many.

Most varieties of tomatoes are indeterminate. They follow the growth pattern associated with vines and continue to grow taller and taller, producing leaves, flowers and fruit at a steady rate throughout the growing season. Indeterminates make great salad tomato plants since you get a good, long supply – little and often. However, this behaviour can spread the plant's energies too thinly. Side shoots appear between the major branches and the stem. Sometimes these shoots are productive, sometimes they are not. The decision whether or not to remove these side shoots has the gardening community divided. Some religiously remove every side shoot on the basis that it concentrate the plant's energies into fruit produced by the main branches. Others leave these shoots and wait to see if fruits develop on them. I sit somewhere in the middle. Depending on how I feel my plant is coping with its pregnancy, I occasionally lighten the load. There is another good reason for removing some or all of the side shoots. If you allow the plant to get too bushy, you run the risk of reducing light to the fruits so they might not ripen as well as you'd like.

Determinate tomato varieties, often called bush tomatoes, behave a bit differently to the vine types. They reach a determined height and go for one big session of fruit production over a period of a few weeks. They are great for producing crops for sauces since you can harvest quite a lot at one time and they have the added advantage of not needing any kind of pruning. This makes bush tomatoes ideal for growing on windowsills, in any kind of container, and even in hanging baskets.

With most fruiting plants, don't be too greedy. When you've got enough flowers on the plant, remove the growing tip. On outdoor tomato plants, around four trusses (little groups of flowers) are usually considered to be enough for any one plant to cope with. Indoors, plants can usually cope with six or seven but it's up to you as to how tall you want to allow your indoor plant to grow. With bush tomatoes, you don't have to remove the tip since the plant will regulate its own height, flower and fruit production.

Regular watering of fruit-bearing plants is essential. Periods of drought cause the fruits to shrink as they dry up, and swell again when they are next watered, splitting the skin of the fruit, so make sure you keep the watering can handy.

Finally, remember the plant is eating for two or more so a little supplementary feed won't go amiss. A high potassium feed will encourage better fruit production – any of the tomato feeds you can buy off the shelf will do for most fruiting plants. A good organic feed is home-made comfrey liquid and you can make this easily by steeping comfrey leaves in water for a few weeks, stirring them every so often. You'll know when it's ready because the infusion will take on its own distinct smell! Dilute your comfrey infusion by around one part comfrey liquid to ten parts water and water into the soil.

Don't feed plants bearing fruit with a high nitrogen plant food as this may stimulate too much leaf production. Feed regularly with a low nitrogen, high potassium feed as soon as you see the fruits begin to appear.

# Giving Birth

When the seeds are fully ripe, a plant's association with its offspring is almost over. Most plants make terrible mothers. Once they have given birth, they have no more interest in their children and compete with them for food and light the same as they would with any other, unrelated plant. For its own survival as much as for the survival of its progeny, a mother plant must do one last thing for its children. It must ensure they have a reasonable chance of obtaining light and food. As you might expect, plants have evolved some ingenious ways to spread their seed far away from the parent plant and even from their own siblings.

Many plants favour wind dispersal. They produce small, light seeds and can achieve seed dispersal by swaying in the breeze. Others have devised methods of flight clever enough to make the aviation industry jealous.

Dandelion seeds are brilliant at flying. With tiny hairs that behave like parachutes, dandelion seeds can travel up to 500 metres. Anyone who's blown on a dandelion head (and who could resist?) will testify to the beautiful simplicity of these flying machines. This ability to disperse seeds far and wide has earned the dandelion a place in the top ten of our most prolific of weeds, doing very nicely without any help at all from humans.

Other seeds have papery wings much like the wings of an aeroplane and glide to the ground some distance from the parent plant. Helicopter seeds spin around to create a vortex that opposes gravity and lengthens the time it takes for the seed to fall to the ground. The longer it takes, the farther the seed travels.

Plants growing close to rivers or seas use water to carry their seeds miles away from their parent and sometimes even across continents. The coconut palm produces a nut containing a single seed which is very buoyant. It has a waterproof coating to protect the seed from the salty sea during its perilous journey.

Plants like Scotch broom *(Cytisus scoparius)* give birth rather more dramatically. Once ripened, the seed pods explode, spewing out seeds over quite a distance. You can sometimes hear a bang or a pop when they do this. The warmth of the sun triggers the explosion. This way the plant ensures the seeds are widely dispersed when it's warm enough for successful germination.

Perhaps the most efficient way to disperse seed is by using other living things. Some seeds hitch a ride on passing animals. They have sticky coatings or burrs that stick to the coats of furry animals and even to human clothing. Cleavers (*Galium aparine*) is a good example. Anyone who's ever had to pull, 'sticky-willies' off their dog or cat's coat will understand how useful this method of transport can be.

Many plants have seeds encased within an edible fruit. This is not an unfortunate accident for the plant. It actually wants its seeds to be eaten. When the time is right, the fruits change colour to signal to other animals they are soft and tasty. Once eaten, the embryo remains unaffected by the digestive system until, sometimes miles away from the parent, the seed is delivered to the ground in the animal's droppings. The passage through the animal weakens the hard outer coating of the seed and allows water and air to enter, initiating germination. This means a parent plant can ensure she doesn't have competing seedlings. Only by being taken away by animals can the seed be encouraged to germinate. Growers sometimes have to soak seeds in sulphuric acid to replicate the acid in the guts of animals.

Squirrels spend autumn collecting nuts and seeds to store over winter. They dig holes all over their territory, hiding the food from others until they are able to dig it up again when times are not so plentiful. However, they don't always remember everywhere they put their stores and some get forgotten. These seeds are then perfectly positioned in the soil with a little help from their gardening buddies.

Once flowering plants have dispersed their seeds, they are exhausted. For annual plants, their whole life's work is over and they wait until they are killed off when the temperature gets low enough. Most perennials (plants that live for more than two years), have just enough time to catch the last of the sun's warmth so they can strengthen their root systems before winter sends them off to sleep.

The little seeds, packed with all the genetic programming they need for them to grow and become parents themselves, will now wait until the time is right for them to join the plant race and ensure the survival of their species for yet another generation.

## Did somebody tell a joke?

We've all seen tumbleweed rolling by in the background, accompanied by the sound of wind, to emphasise the awkward silence that follows the telling of a bad joke. But for the Russian thistle, thought to have arrived in North America on a shipment of flaxseed, it's no laughing matter. Russian thistle - *Salsola kaliis* – makes the ultimate sacrifice for its children. Its final act is to commit suicide to spread its seeds.

When the time comes, the tumbleweed

simply stops living and dries up until it becomes detached from its root system. The dead plant, pushed along by the wind, carries its children with it and, with each jolt and bump over stony ground, disperses its seeds along the path it takes. For the Russian thistle, its kamikaze mission is well worth it since each plant, depending on how big it is, can disperse up to 100,000 seeds.

## How to stop those 'evil' weeds

We might think that weeds are evil things, bent on making our lives a misery but they're not. In fact, as someone who loves plants for plants sake, I have a special place in my heart for weeds, so much so, my garden is full of them. I admire weeds because they have superior survival skills and many of them manage this by being super-efficient at seed dispersal.

If you want to reduce the amount of weeds in your garden, the best way to do this is to prevent them from seeding altogether. There's really no need to get down on your hands and knees and pull them out by hand. As long as you're prepared to do it regularly, you can stop them in their reproductive tracks by taking a long-handled hoe and slicing off the flower heads before they have a chance to produce seeds.

Dead heading in this way has to be done on a regular basis – every week if you can – but it's not too hard a job if you've got a sharp hoe. If you get your hoe right down to the base of the plant and remove the whole top growth, you will increase the amount of time it takes for the plant to recover and produce a flower again.

If you haven't caught your weed in time and you remove it after seeds have been produced, then don't put it in the compost heap. The seeds will probably survive the composting process and, when you spread the rich compost over your garden the following season, you'll be sowing a number of weeds as well.

## Postnatal care

After ensuring the survival of yet another generation, perennial plants turn their attentions to their roots. Exhausted by the rigours of pregnancy, they're planning on having a rest but, before they do, they must strengthen their root systems to gain a head start when they wake up for a new growing season. You can help your plant at this time by feeding it with a balanced fertiliser. Avoid a high nitrogen fertiliser because this will produce too much green growth and would only be a waste to the plant so close to its rest period. Better than applying a chemical fertiliser, mulch the base of the plant with homemade compost. The compost will release nutrients slowly and keep the soil at a more consistent temperature throughout hibernation.

After flowering, container plants may have exhausted the nutrients within the pot and the roots have nowhere else to go to search for more. Feed your container plants regularly and don't forget about them as winter approaches. Some fresh compost applied to the top of the container will give them a new supply of nutrients to help them cope with the oncoming winter.

# Part Three: A New Life Begins

# Waiting for the Right Conditions

No one knows exactly how many individual plants there are in the world. Plants are born, live, die and get eaten every day, so counting them would be an impossible task. The estimated biomass of plants is around two thousand million tonnes. That's a whopping 20 times the weight of all the humans. We can assume then that several million, probably even several billion plants are born every day. Mostly, but not always, they come into being through germination. Germination is the process by which a dormant seed wakes up and begins to sprout. All over the world, every day, every minute, every hundredth of a second, a new plant is born and this remarkable feat is achieved without so much as a second look from most of us.

When a seed hits soil, it must wait for the right conditions to wake up and begin its new life. Unfortunately for some plants, the window of opportunity is just a few days but, for others, it can be more than fifty years. Little chickweed seeds can hang around for forty years or more, hence the reason why chickweed is one of our most common weeds.

Most of us know all we have to do is

introduce our seed to soil, cover it up a bit, water it, and some days or weeks later, a new plant emerges. Simple really, but if we want to be really successful at growing plants, it's good to know exactly how this happens.

Inside its hard outer shell, the seed of a typical flowering plant has a tiny little embryo with either one or two seed leaves. On the seed coat itself, is a scar – the hilum – the seed's very own belly button, which tells us it was once attached to its mother's ovary. A minute pore marks the point at which the pollen tube entered the ovule during fertilisation and it is through this pore that water enters. Our seed needs just the right amount of water to break its outer shell and start it into growth. Water causes the seed to swell and the shell is broken.

It wouldn't be a good idea to germinate just as winter approaches since the young plants might not cope with the fall in temperature. Most of the plants we grow germinate when the soil temperature is between 15 and 23 degrees Celsius. Some germinate at lower temperatures, some at higher. To prevent accidental germination during an unusually warm spell in autumn, many plants have seed coats so hard they have to be softened over winter by colder temperatures. Water alone won't be enough to break the coating even when the temperature is right. They need this cold period to break them out of dormancy.

Even at the right temperature and with the correct amount of moisture, germination will not occur without oxygen. The mother has left her seed with a little gift of nutrients to provide enough energy for the seedling to haul itself into the light. Without oxygen, the seed cannot process this stored energy. If the soil is heavy and waterlogged, all the air spaces are pushed out and there won't be enough oxygen. So seeds need to be damp but not too wet or they will fall at the first hurdle.

The final piece of the germination jigsaw is light. Some seeds need darkness for germination; some won't germinate without sufficient light. Insufficient light levels occur when the seed is buried too deep or if there is a lack of space in the canopy of plants above. Seeds can wait until more light becomes available. When a tree falls, there is a sudden gap in the canopy and many seeds will use this as a signal to spring into life.

But how is it possible for a seed to sense light at all? Light in the red wavelength affects phytochrome, a plant pigment within the seed. Only when a seed finds itself at the correct depth will the phytochromes stimulate germination. If you've ever wondered why, when you dig over an area of soil, it seems to produce an explosion of weeds, this is because you've brought some light-sensitive seeds closer to the surface. Sensing they are nearer to light, the seeds break dormancy and sprout.

So seeds can wait, some even for hundreds of years, until conditions are favourable. They need all four pieces of the jigsaw to be really successful – moisture, temperature, oxygen and the right level of light.

## All alone in this world

Over two thousand years ago, a group of Sicarii Jews were besieged by Romans, trapped in their fortress at Masada in the Judean Desert. To help them through the siege, they stockpiled food in the fortress' storeroom. By the time the Romans finally breached the fortress, the Jews had destroyed everything in the fortress and committed suicide rather than be taken. Legend has it their leader gave orders for the food store to be left intact so the Romans would know the Jews had chosen to take their own lives, even though they had plentiful food supplies. Amongst the foodstuffs to escape the initial destruction of the fortress were dates, believed by the people of the time to have both medicinal and nutritional value. Inside the fruits were the seeds of the Judean date palm.

For two thousand years, like the boy-robot in *Artificial Intelligence*, one date palm seed waited in the ruins of the fortress while the world above moved on and the little seed's closest relatives – the cultivar to which it belonged - became extinct. In 2005, when the Blue Fairy finally arrived in the shape of plant specialist Elaine Solowey, this seed was the only one that successfully germinated, making it the oldest directly dated tree seed to be germinated. They called the plant Methuselah.

Date palm trees are dioecious and it was six years before Methuselah finally flowered and announced to the world that he was indeed male. He was then introduced to a modern female date palm and proved his virility by becoming a father. With its legendary healing properties, Methuselah may hold the key to curing illnesses that weren't even around when his parents were alive and now the search is on for a viable seed from his own era to provide him with the perfect mate.

# Why do I have to put my seeds in the fridge?

Many perennials, most hardy trees and shrubs, and carnivorous seedlings, need a period of cold before they can break out of their tough seed coats. If you buy seeds requiring this, it usually says so in the instructions on the back of the packet. If you see the term 'cold stratification', then you know you need to pre-treat your seeds.

You could just plant these seeds outside in autumn. However, so you have complete control over the whereabouts of your seeds and their germination, you can place your seeds inside a plastic bag, add some dampened vermiculite or sand, seal the bag and put it in the fridge for the required amount of time – usually a couple of months. For very small seeds, it's best to sow them in a seed tray before you put them in the fridge.

Be sure to label and date the bag so you can take it out after the proper time. Once out of the fridge, sow the seeds in the usual way at room temperature. As a general rule, almost all of our native tree seeds will appreciate this kind of pre-treatment. For seeds which stubbornly refuse to germinate, you might have to break the seed coat manually by nicking or scoring it with a knife or rubbing it with a little sandpaper.

## How deep do you need to go?

Seeds are both a potential new life and a food store. The smaller the seed, the smaller the food store and the more important it is for the emerging plant to reach the light before its stores are exhausted. If it takes too long for the plant to reach light, it will run out of food before it gets there.

In the wild, seeds get buried naturally by passing animals, by rain washing soil over them or in any number of ways. When a seed gets surrounded by soil, it helps to retain moisture, so most seeds like to be buried. If a seed gets buried too deeply for its size, it may fail or it may, like chickweed, have the ability to wait until it's brought closer to the surface again. Larger seeds, usually because they are not in contact with sufficient all-round moisture from the soil, may lie dormant on the surface until they get buried deep enough.

A major reason for failure when we attempt to grow seeds is that we plant them at the wrong depth. The instructions on the seed packet will tell you the correct depth for your seeds but, if you don't have instructions, then the depth should be around three times the width of the seed. For very tiny seeds, this can sometimes mean simply scattering them on the surface and not covering them up at all.

## Step by step seed sowing

**Step one:** Prepare your seeds, if you have to, by placing them in the fridge or by nicking with a knife.

**Step two:** Fill your seed tray with a suitable growing medium – vermiculite or seed and cuttings compost is best. **Water** the growing medium, but don't soak it. It should be like a wrung out sponge. Remember, too wet and the seeds won't get enough **oxygen**.

**Step three:** Spread the seeds over the growing medium – the bigger the seeds, the wider the gaps between them. Cover the seeds with soil or vermiculite to lock in moisture. Don't water again for a while or you will risk dislodging the seeds when they're at their most vulnerable. Instead, place the seed tray in a closed case propagator (one with a clear lid) or cover the tray with cling film to prevent moisture from escaping.

**Step four:** Place the seed tray on a sunny windowsill at room **temperature**. This ensures the seeds have immediate access to **light** as soon as they appear.

**Step five:** Wait and watch. Check moisture levels by placing your hand on the growing medium. If it feels a bit dry, use a little hand mister to gently spray it with water.

That's it! Couldn't be easier. There's nothing else for you to do now until your seedlings emerge.

# Abandoned at Birth

Plants are completely independent from the moment they are born and nobody teaches them what to do or protects them from harm. They come into this world equipped with all the genetic programming they need to survive into adulthood. With their roots programmed to head downwards and their shoots programmed to head up towards the light, almost every plant occupies two very different spaces at one time. The roots spend their lives in darkness, seeking out water and dissolved nutrients while the stems, leaves and flowers exist in a world of energy-giving light. Immediately after germination, the seed is transformed from a dormant seed into a seedling, ready to fulfil the wishes of its parents and ensure the continued survival of its own species.

Plant babies are born feet first. The root follows gravity and anchors the plant in the place where it will spend the rest of its life. With the stored food source in the seed all that stands between life and death, the root must act fast and seek the water and nutrients it needs to support the stem. The stem emerges soon after and it must defy gravity and push up through the dark soil until it reaches the light. In some plants, at the top of the newly emerged shoot are the seed leaves, unlike the plant's eventual leaves because they are formed within the seed with the primary purpose of providing energy quickly. They will usually be curled into a hook to help the stem push through the soil. Sometimes the leaves will hold onto the hard seed case to further assist its passage through the soil. You can see this in bean plants where the seed case is clearly visible at the top of the emerging seedling. In other plants, like sweet corn, the cotyledons (rudimentary leaves inside the seed) stay below ground.

Once through the soil, the stem shrugs off its seed casing before opening its first set of leaves. Then, the plant begins to draw further on the food stored within these leaves. If the plant is a monocotyledon – possessing just one seed leaf – it will have to produce another leaf as quickly as possible. If the plant is a dicotyledon – producing not one, but two seed leaves - then it is able to perform a miracle immediately. That miracle, the basis for all life on this Earth, is known as photosynthesis.

The plant's first act is to collect water via the roots. Using energy from the sun absorbed by its tiny leaves, it breaks the water molecules down into hydrogen and oxygen. Having no need of all the oxygen at this stage, it releases some of it back into the atmosphere. Next, it plucks carbon dioxide from the air and combines it with hydrogen to produce sugar which it will use as a food source. In essence, this is the formula for photosynthesis. Water plus carbon dioxide, using energy from the sun is converted into sugars plus oxygen. The balanced equation for all this is: $6H_2O + 6CO_2 +$ sun energy $= C_6 H_{12} O_6 + 6O_2$.

For the little seedling, it is imperative it is able to perform the miracle of photosynthesis as soon as possible so the seed leaves – the seedling's first set of tiny solar panels - are relatively simple. They are pretty plain affairs and often bear no resemblance to the leaves of the adult plant. The leaves of courgette (zucchini) plants are palmate (shaped a little like the palm of your hand) with fine hairs which can irritate the skin. Courgettes are loved by slugs and the plants have probably developed these fine hairs to irritate the soft skins of slugs, but it takes a lot of time and energy to produce such complexity and the first set of leaves on a courgette plant are plain and oval. These simple leaves will do the job of gathering sufficient energy to produce the plant's 'true' leaves.

Unfortunately, for many seedlings, the formation of the seed leaves marks the beginning and end of life. With only a small root to anchor them to the ground, seedlings are at risk of being trampled, washed away in heavy rains, eaten or cut off by a gardener's hoe. Those that survive the first few days of life are the lucky few.

If our plant does manage to survive this crucial time, underground, the primary root delves deeper, increasing the plant's chances of holding fast when disturbed and sucking up whatever moisture and nutrients it can find. Eventually, the root system expands laterally, stabilising the plant further. Root hairs increase the surface area of the root zone and improve the chances of making contact with water. Near the tips of the roots are neurotransmitters. No one knows yet exactly how they are used, but neurotransmitters are chemical communicators and are most likely used to communicate environmental stress. Above ground, the seedling listens to its root system and develops in tandem with the roots so it can achieve height and stability. Both above and below ground, there are meristematic cells. Below ground, the RAMs (Root Apical Meristems) near the tip of the roots is responsible for lengthening the root. At the tip of the shoot, the SAMs (Shoot Apical Meristems) lengthen the stems.

Above ground, the primary SAM continues to stretch towards the light and the plant produces its first set of true leaves. These leaves will be the same shape as the leaves on a fully developed plant. Tiny as it is, the seedling has now become identifiable as the offspring of its parents and the next generation of its species.

## Sunflowers at play?

Do young plants indulge in playtime? Professor Stepano Mancuso thinks they might. He claims that time-lapse photography of young sunflower seedlings shows movement which cannot be attributed to these seedlings simply following the pattern of light. He thinks these sunflower seedlings may actually be playing – practising the movements they will have to make as adults and strengthening their stems in the process. This kind of play would be similar to that of young animals who might practise catching prey by rolling around and pouncing on their siblings. It's an interesting thought and I'm going to sit on the fence for now. However, I'd really love it to be true....

## Bringing up baby: caring for young seedlings

When your seedlings first emerge, they're going to need your help, especially if they're trapped in a seed tray or pot. Arguably, this is the stage of their lives where plants gain the most advantage from their friendships with humans. For these vulnerable little babies, being adopted by a human is a definite plus.

The most important thing they'll need is moisture. However, having gone to all that trouble to push up through the soil, the last thing they want is to be flattened by a torrent so don't water your seedlings from the top unless you use a very fine spray mist. It's better to water them from below by standing the seed tray in water so the roots can do their job. Only stand them in water long enough for the growing media to take up moisture and don't leave them standing in water too long or you'll risk drowning the roots. A good way to keep your seedlings just moist enough is to sit them on capillary matting or wet newspaper. If you use this method, ensure both the growing media and the matting are wet to begin with otherwise the matting will do the opposite of what you want and draw water out of the compost.

If your seedlings are outside, environmental factors will have made them stronger than your indoor ones, but they still need to be watered with care. Use a fine rose on your watering can, start off the flow of water away from the seedlings and bring the water towards them once it's in full and even flow. Hold your watering can close to the plants to reduce the impact of the water.

If you haven't sown your seeds far enough apart, you're going to have to carry out an early cull. This is the bit I hate because I can't help feeling sorry for the little ones I throw away, but it has to be done if you want the survivors to get enough light. Simply pull out some of the seedlings and discard them. You can carry out this task soon after the seedlings appear. The ones you choose to survive will thank you in the long run. This process is known as 'thinning'. Thinning can be less severe for indoor seedlings if you intend to transplant them later into bigger pots but, for seeds sown outdoors, you have to thin in the early stages. Increase the space between the plants to around a quarter the distance recommended on the seed packet. Later, you need to thin again, choosing to keep the healthiest plants and discarding the weaker ones until you have the correct spacing.

You must ensure your seedlings are in sufficient light. Thinning and discarding some of the seedlings will help. If you've got them on a windowsill, you might find they'll lean towards the light. You can help them to grow straight by turning the tray every now and then so they have to lean the other way and straighten up. If your seedlings have to stretch too far, the stems won't have time to thicken and get stronger. If you're growing your seedlings indoors and you see them getting long and leggy, you might have to give them some artificial light. The artificial light source should be close enough to make a difference, but not so close you burn the leaves.

You don't need to feed them at this stage.

There should be enough nutrients in the growing medium and you don't want to encourage them to grow too fast, too quickly. What you want is short, stocky stems rather than long and thin ones.

For indoors seedlings, provide them with good air circulation. Indoor seedlings can succumb to a fungal infection called 'damping off'. The fungus attacks the seedlings at the base of the stem and they appear to keel over for no apparent reason. Thinning helps, as does allowing the soil to dry out just a little before watering again, but good air circulation is the key.

Outdoors, your seedlings get tossed around by the wind and have to cope with being pushed over from time to time in the rain. All this movement is good exercise for the seedlings and, though they may grow a little slower than seedlings indoors, they will be stronger and stockier. To toughen up your indoor seedlings brush them regularly with your hand or fan with a newspaper.

When your seedling has developed around three sets of true leaves, it's telling you its root system can cope with being moved, either into a bigger pot or outside to its final growing position. This process is known as 'pricking out'. A young plant judges the proximity of others by sensing light levels and the shadows cast by other plants and nearby objects. If there isn't enough room, it may not grow as quickly as you want it to. Moving plants into bigger pots lets them know they've got room to grow and they'll usually get going faster when you do this. If you're pricking out into individual pots, choose a fairly small one. If you choose too big a pot, then you risk losing parts of the roots when you transplant later because they are too loosely spaced and may break easily.

Fill your small pot with a good medium, designed for potting on. Carefully loosen the potting medium around the root and lift the seedling from the seed tray taking care to keep its root system intact. Holding the plant gently by the leaves, make a hole in the new compost to accommodate the plant to the depth it was at before. Place it in the hole and firm the soil around it so you have no gaps. Gaps in the potting medium around newly transplanted seedlings can fill with water and drown the plant. Don't move the plant again until you're sure its root system has just about filled the pot.

If the young plants are going directly outside, they will need to be 'hardened off'. This means you must get them used to the temperatures outside without sending them into a state of shock. Put your plants outside during the day and take them in at night. Do this for a week or so and it won't be such a trauma when they experience outdoor night temperatures for the first time. Hardening off should be done for all plants grown indoors with the intention of planting them outside.

Failure to harden off plants can result in transplant shock. This severely checks the growth of the plant and, in the worst cases, kills it altogether. When you take a plant out of a warm greenhouse you trigger a response similar to that which occurs in autumn. The plant feels a sudden, noticeable drop in temperature and, thinking winter is approaching, begins to enter a state of dormancy from which it will take some time to recover. If you buy bedding plants from a supermarket or large DIY store (and I'm not necessarily recommending you do this) it's possible they may have been mass-produced and delivered direct from the greenhouse so you usually have to harden them off. In most cases, if you go to a reputable local nursery, they will already have done this for you.

# The Pack Leader Emerges

Nearly all plants develop both above and below ground. Above ground, the stems, leaves and eventually flowers establish the shape and form of the plant. They are the face of the plant – the part we recognise – but, unseen by us, the roots are equally as important.

Below ground, the roots must push through the soil, send out anchors to stabilise the plant, and search for food and water. However, the root system must have sugars to give it energy. Without the stems and leaves, the roots would starve and without the roots, the stems and leaves would not have adequate anchorage, water and soil nutrients to stay upright and produce leaves and secondary stems.

In the early life of a plant, it is the primary shoot's job to ensure sufficient energy is manufactured and it must do this as quickly as possible so it can compete for sunlight with all the other plants around it. To do this, in some plants, the first of the shoots to appear establishes dominance over all the others. It is, if you like, the pack leader.

As the dominant shoot apical meristem

(SAM for short) pushes towards the light, it deposits little bundles of cells (secondary meristems) along the stem from which new leaves will emerge but, as they are yet undefined, they also have the potential to produce side shoots. A dominant SAM at the very tip of the main shoot, exerts control over a hormone that inhibits the growth of these new side shoots until it can reach up high enough to receive sufficient sunlight. Like any other pack leader, it is saying to the rest of its pack "No one gets to eat until I eat first".

When the dominant SAM draws away from the other potential branches, its hormonal influence over them weakens, they are allowed to break into growth and the plant produces branches with subordinate SAMs. Because of the influence of their pack leader, these branches develop low enough down the stem so they don't overshadow the dominant SAM and compete with it for light.

As the leaves above manufacture more sugars through photosynthesis the root is able to use these sugars to produce lateral roots below ground to feed the development of new branches and provide extra anchors. Just as some plants have a dominant SAM, so some plants have a dominant RAM. Where there is a dominant RAM, the root system consists of an autocratic, central tap root with subordinate, lateral branches. In plants with no dominant RAM, the root that first appeared from the seed withers away to be replaced by a democratic government of fibrous roots of equal importance. Whether the plant has a dominant tap root or a number of fibrous roots, these roots will, in turn, produce subordinate branches to further assist the plant in its search for water and nutrients.

In no other group of plants is the ever-dominant presence of the SAM pack leader more evident than in the trees. Looking at a tree, it is easy for us to see the dominant stem with its bundle of leaves at the very top, always in the best position to achieve the most sunlight and the less dominant, branching stems arranged below.

But what happens when a plant loses its pack leader? Just like in any other pack, there are always leaders-in-waiting and for most plants, these leaders-in-waiting will be the SAMs in the next strongest position - in other words, the branches nearest the top. If the dominant tip of a plant is lost before there are any established branches below, new stems will develop very quickly at the next bundle of cells down, usually where there is already a set of leaves. If there are already branches, one or more of these branches establish dominance and grow faster with the new growth becoming more vertical.

In almost all cases, when the dominant shoot apical meristem is lost, the plant becomes bushier as one or more of the lower branches compete to achieve the top position. It is by understanding the phenomenon of apical dominance that we can solve a major gardening mystery – when and what to prune. Once you understand what the plant will do when you remove its pack leader, you will never be afraid to tackle pruning again.

# Pinching out bedding plants for increased flowering

Left to their own devices many bedding plants follow the growth pattern associated with apical dominance. There will be a tall central stem and weaker lateral branches, all of which produce flowers. But, if you want a bushy plant producing abundant flowers all over it, you need to remove the pack leader.

When the main stem has produced a few sets of leaves, and before it gets too tall, follow the main stem from the top to the next set of leaves down. With your thumb and index finger, break the main stem tip off. This process is known as 'pinching out'. In no time at all, the plant will produce at least two lateral stems at the point where you removed the tip.

These new pack leaders begin to grow vertically so, after they have produced a couple of bundles of leaves each, remove these new pack leaders and you'll get further branching. Since you have removed two competing pack leaders, you'll get four or more in their place, thus encouraging the plant to become bushier.

Every time you remove the tips of the branches, you delay flowering so you don't want to do it indefinitely but, if you pinch out three or four times in the early stages, you'll get a shorter, bushier plant with flowers spread throughout the plant.

## Pruning shrubs

The idea of pruning shrubs can fill many new gardeners with horror. When is the right time to prune? Where do I make the cut? Will I kill the plant if I get it wrong? Gardening books don't always help. They say things like: *'prune plants that flower on old wood in summer'*; *'prune plants that flower on new wood in spring'*; *'cut to an outward facing bud'*; *'prune just above the node'*. I sometimes think they make it all seem too complicated so the experts can appear to be members of an exclusive club while the rest of us remain in the dark.

The truth is, pruning shrubs is easy and you don't have to know the name of the shrub you're pruning to get it right. Just take a good look at its growth habit and it'll tell you everything you need to know. Let's start with **when** to prune.

Almost all shrubs are perennial and, as

such, they need to live through winter. As the temperature drops and the days shorten, all plants - even the evergreens – take a rest in one way or another to conserve energy. For evergreens, their growth slows drastically, but they still retain most of their leaves. Others shed their leaves altogether and enter a state of semi-hibernation.

Spring-flowering shrubs have only a short window to establish the buds that will eventually become flowers so they plan ahead by forming their flower buds before they go to sleep. Imagine you have to get up really early for an important appointment. Most of us would lay our clothes out the night before because we would know we wouldn't have enough time in the morning to think about what we're going to wear. That's exactly what both evergreen and deciduous spring flowering shrubs do before they bed down for winter. Their flower buds are all neatly folded up, laid out on the branches of the wood they formed during the growing season.

If you prune spring flowering shrubs too late in the year or early in spring, you remove buds which were intended to produce flowers and you reduce the plants' potential to flower for that year. Spring flowering shrubs should be pruned in early summer, just after they have finished flowering, and before they begin to plan next year's flowers. This is where the expression 'flowering on old wood' comes from. It simply means the flower buds are formed in the previous year's growing season and before winter sets in.

Summer-flowering shrubs have enough time after they wake up in spring to plan for flowering. Just as spring approaches is the best time to prune them. In other words, just before they begin to form new buds. These shrubs are said to flower on 'new wood' because, before they dress themselves in flowers, they have time to produce fresh new growth on the stems and branches. If you prune these shrubs in early spring, you can prune out any damage done over winter and control the direction of the new growth and, ultimately, where the flowers will be.

Flowering shrubs with a tendency to become leggy (bare at the bottom with all the flowers at the top) such as lavender, heaths and heathers should be pruned religiously every year as soon as they have finished flowering. So, for many, the time to prune them will be autumn. Give them a light trim without cutting into old growth.

And what will happen if you prune flowering shrubs at the wrong time? Don't worry too much about that, it's highly unlikely you'll kill them. You might upset them a bit by ruining their flowering plans for a whole year, but they'll get over it. Prune at the wrong time and the worst you'll get is a poor display of flowers for that year and a slightly annoyed plant.

For shrubs you're growing for foliage, you can prune them in late spring to early summer and, for hedging plants, a light prune once or twice during the growing season will keep their branches fighting for dominance and retain a nice thick growth. As a general rule, it's not really a good idea to prune shrubs in late autumn or early winter as this exposes the cuts to frost and can damage them.

Now we look at **how** to prune shrubs. Where possible, you need to locate the nodes on the stems. These are the locations of the little bundles of undefined cells the SAMs have left behind. The nodes are slightly fatter growths on the stems and, in most shrubs, you should find them easily. Often, there will be a lateral bud at the node which will ultimately break into a branch. If a lateral branch has already formed there, check you are happy with the direction it's taking because it'll probably take a stretch as a result of pruning. The part of the stem between the nodes or lateral branches is known as the internode. Internodes don't have the potential for new growth so, if you leave an internode on the stem, it will usually 'die back' to the nearest node. This results in blackening of the cut stems which is unsightly and can encourage disease to enter through the wound. Make your cuts just above the node to allow the plant to burst forth from the point at which it was cut.

You prune shrubs for health, height and shape. So we start with health. Examine the shrub really closely. Check for diseased or damaged branches and remove them. Check also for weak and spindly branches and take them out too. Finally, any branches rubbing against one another will cause wounding so get rid of these.

Remember, pruning is plant surgery and you'd never see a surgeon use the same scalpel on two different patients without sterilising it in between. If you want to reduce the risk of cross-infection, wipe down the blades of your pruning tools with disinfectant before you move on to another plant.

Next, we prune for height. Each time you cut the main stem it temporarily suspends upward growth. However, the new pack leaders will begin to grow vertically, so you need to keep each subsequent set of dominant branches under control every year. Pruning for height is really a question of personal taste and location of the plant. Prune for height if your shrub might obscure a window or other feature and also with flowering shrubs if you don't want all the flowers to be at the top with little or no flowers underneath.

Finally, we prune for shape. We know many plants have a dominant stem and less dominant branches. These less dominant branches also have a hierarchy so they have even less dominant branches growing outwards from them. Each time you prune a more dominant stem, the stems below break into dominance and so on. Knowing this can allow you to anticipate exactly how the shrub will branch, and in what direction.

For shrubs with nodes and leaves arranged directly opposite one another, for every stem you prune out, you'll get two in their place. For those which are arranged alternately – i.e. one bud facing one way and another lower down, facing the other way – you can control the direction of the new branch by pruning to a bud that faces outwards. This way you prevent a new stem from growing into the middle of the shrub where it will bump into and become entangled in other branches. That is why you are often advised to 'prune to an outward facing bud'.

In cane-forming or multi-stemmed shrubs such as dogwoods, the dominant stem isn't always evident since branching occurs so far down that all we see is a collection of stems which appear to come directly from the ground. These shrubs should be pruned by cutting one in every three stems (tallest first) to ground level. This forces it to produce fresh growth from the bottom and the shrub will keep its unique shape. Each year, you cut out the oldest stems so each stem regenerates once every three years. This keeps the shrub nice and open and maintains its youthful vigour.

In all cases, pruning should be undertaken with sharp tools so you get a clean cut and the cuts should always be made just above a node or bud.

In the case of conical shrubs and conifers where you want to retain that Christmas-tree shape, you should never remove the dominant SAM as you will probably lose that shape forever. These conifers should be trimmed lightly, by brushing the sides with hedge trimmers or hand shears. In the case of almost all conifers, avoid cutting beyond the green growth. Cutting into the brown branches of conifers will result in a tree that is bald, if not forever, for a very long time.

# Being a Good 'Mother'

In evolutionary terms, sexual reproduction is by far the most preferred method since the mixing up of genes between two plants is likely to result in a stronger plant. However, many plants are capable of producing new plants without indulging in sex at all.

Some plants send out long stems (runners) which form new roots and stems once they make contact with the soil. The resultant baby plants (or plantlets) remain attached to the parent plant, receiving nutrients and moisture until they are strong enough to grow independently. It may seem like the parent plant is being a good mother, tending to her nursery of plantlets but it's not really a mother since the plantlets it produces are exact clones. These 'new' plants will be the same age as the parent and possess exactly the same genes. They are an insurance policy – a way for the plant to occupy a new piece of ground in case the one it's currently in isn't quite up to the job. One of our most popular houseplants, the spider plant, *Chlorophytum comosum,* produces long runners with 'baby' plants at the end, as do strawberry plants and many species of grasses. Spider and strawberry plants have above ground runners, but other plants have runners below ground. These runners are often confused with roots and produce new plants at regular intervals, thus allowing a single plant that cannot increase in height, to increase in size by expanding outwards.

Generally, when a plant produces runners above ground, they are known as stolons. If below ground, they are rhizomes but the difference between stolons and rhizomes is actually a little more clearly defined. Both stolons and rhizomes are stems, not roots. The real difference between the two is that stolons are produced from the main stem, but rhizomes actually *are* the main stem. Think of a grassy plant with only leaves and flowers above ground and no noticeable stems and you'll know what I mean. These plants have stems which allow them to expand, not so much upwards but sideways. You just can't see them.

Often, after a time, runners wither away or are otherwise destroyed and the new plants are able to stand alone. Many rhizomatous plants can withstand fires, being eaten, and all sorts of attacks because their stems are protected underground. Each cloned plant is capable of sexual reproduction or producing runners of its own.

## Bramble: the mother of all plants

When people ask me to name my favourite

plant, they are surprised to learn it's the wild bramble *(Rubus fruticosus)* a plant considered by most gardeners to be nothing more than an annoying weed. But for sheer tenacity and survival skills, you've really got to admire the bramble. This plant is deadly serious, not only about ensuring its survival, but about spreading as far as it can.

It creates thorns as protection from grazing animals, buying itself the time it needs to grow a main root up to four metres in length. Its stems can travel laterally for around five metres. In the meantime, it's courting as many species of wildlife it can, by producing masses of pollen-rich flowers which will eventually become edible fruits. The fruits are eaten by birds, other animals and even humans, thus transporting the seeds over a wide distance. But the bramble doesn't stop there. This plant also produces clones.

Stems of the bramble push out in all directions, clambering over other plants, walls and fences. After up to five metres, the stem drops to the ground where it forms roots. The new plant sends out its own long stems and, here we go again, another five metres travelled.

A wild bramble plant isn't just a good mother to her own clones; she acts as a surrogate mum to all kinds of wildlife. Her flowers are rich in pollen and nectar and are visited by bees, wasps, butterflies, moths, lacewings and hoverflies. Many insects hide from predators behind her thorny defence and, for the most part, she makes them welcome. Just in case they try to take a bite, she protects her own leaves with a strong, waxy coating and it is this waxy coating that renders her pretty impervious to our weed killers. Hundreds of birds and animals can increase their energy reserves for winter by eating her fruit. Because she is able to form such an extensive, impenetrable thicket, small animals and birds know they can escape larger predators by taking refuge under her protection.

Wild brambles can survive in the most inhospitable of soils and colonise waste ground at an alarming rate and its thorny defence system means we are loath to attack it without proper protection. Fortunately, we have a bittersweet relationship with this highly successful weed. While we may not want it in our gardens, we love to pick its fruit and don't mind it all so long as it's not in our backyard. If you want to pick the fruit, wear gloves to protect yourself from thorns. Watch out for signs of disease on the plant and avoid any that don't look to be in top condition. Always leave a few fruits for the birds. Bear in mind the age-old tip for picking brambles – always pick your fruit from a part of the bush that's higher than a dog can lift its leg!

# Who wants to live forever?

We are often seduced by the idea it may be possible to do away with death one day and become immortal, but it could be some plants have already achieved this. Pando is a clonal colony of male quaking aspen trees who is arguably credited with being the heaviest known organism in the world. He spreads by rhizomes and has reached a size of around 106 acres. Each new shoot he produces looks like an entirely new tree but they're really just another one of his arms – exact genetic copies of himself. Because these 'arms' can die back to be replaced by new ones, it is difficult to age him and some believe the official estimate of 80,000 years is way out. It might be that aspens like Pando can live for a million years or more. Essentially, they are capable of immortality unless struck down by disease or other environmental stresses.

It's just as well that Pando is immortal because it is thought he hasn't flowered for around 10,000 years. Due to climate changes since he was born, it's possible he may never flower again.

# How to employ your own plant babysitters

You can follow the example of the bramble and others by encouraging some plants to produce plantlets. That way, you get new plants with a free babysitter. With strawberries, spider plants, blackberries and some grasses, you don't have to do a thing. Simply wait for them to produce new plants, then remove and pot on the young clones when they're ready. Other plants may need a little persuasion to produce live clones.

Shrubs with flexible young stems such as *Clematis*, lilac, *Magnolia*, *Rhododendron*, *Viburnum* and *Wisteria* can all be persuaded to produce new plants which are fed for a time by the parent plant. This method of propagation is called layering and you can try it with any plant with a stem long and flexible enough. Here's how it's done:

**Step one:** Choose a long, flexible, non-flowering stem and strip away the leaves from the part of the stem that makes contact with the ground when you bend it.

**Step two:** With a sharp, sterilised knife, make a slanting cut just a little way into the stem exposing the softer tissue inside. The best place to make this cut is at a node because this is where the most undefined cells are concentrated. Push a matchstick or cocktail stick into the wound to keep the soft tissue exposed to the soil.

**Step three:** Taking care not to break the stem at the point where you have made the cut, carefully bend it over so the wound makes contact with the soil.

**Step four:** Pin the stem to the ground with a bent piece of wire and cover with around 10cm of soil.

**Step five:** No more care is required except to keep the soil around the layered plant well-watered and wait for the parent plant to do its job.

For most shrubs, it normally takes a full season to produce a completely independent new plant, so a little patience is required. Remember, you are creating a clone, so it is an exact copy of the parent plant in all ways and will inherit any weaknesses. Make sure the plant you choose to be the parent is strong and healthy.

For plants whose stems are not flexible enough to bend to the ground, you can have a go at air-layering. This method is used for houseplants such as rubber plants and for citrus trees, figs, *Camellias* and *Azaleas*. Trim off any branches and leaves from a 30cm section of the stem and cut a 2.5 cm upwards-facing tongue into a node. Stuff moist sphagnum moss into the wound to keep it open and then wrap more sphagnum moss around the whole wound. Cover the area with a plastic bag and secure the bag to the stem above and below the wound. After a while roots will appear. When enough roots have developed, cut the stem away from the mother and re-pot the new plant.

## Mix grass types for the perfect lawn

Some grasses produce runners and some do not. Bunch grasses don't readily produce runners. Instead, they have a single stem which produces thick bunches of leaves from just below soil level. Bunch grasses like perennial rye-grass are popular in lawns because they are extremely hard wearing and respond well to mowing. However, because they don't produce good runners, they can't quickly colonise any bare patches. Rhizomatous grasses are slower to establish and not so good at thickening up, but their rhizomes spread into any vacant ground, filling up bare patches. Most good lawn seed mixes include both bunch grasses and rhizomatous grasses so we get the best of both worlds in our established lawns. Every time we mow, the bunch grasses respond by thickening up and, if any bare patches appear in our lawn, the rhizomatous grasses are quick to fill them. Always check you have a lawn mix that contains both types. This is especially important if your lawn is subject to heavy wear.

The necessary presence of rhizomatous

grasses means we need to carry out the back breaking work of scarification. Scarification is the removal of dead or excessive amounts of rhizomes which, if left unchecked, can clog up the soil surface restricting light and air. Scarifying once a year in early autumn can make your lawn the envy of your neighbours and I think it's gentler and less stressful to the grass plants to use a lawn rake. You can use a powered scarifier but it can be so tough on the lawn you might have to over-seed it.

# Dolly the Sheep ain't so Great

When the first cloned mammal was successfully created, she became the most famous sheep in the world but, if plants could talk, they would have looked at one another and said "So what? Dolly the Sheep ain't so great".

Runners are not the only way plants can produce clones. Plants with underground storage organs such as bulbs, corms or tubers also create clones. In the case of bulbs, little bulbs grow around the main bulb. These 'bulblets' result in new plants but, since they are not formed by sexual reproduction, they are clones rather than new individuals. Of course, potatoes sprout new plants from eyes on the tubers and, once again, these are clones. We may call them 'seed potatoes' but they are not true seeds.

Quite remarkably, some dandelions can manufacture seeds without using pollen. In other words, the mother plant has no need of a father and instead produces exact clones, only in seed form. Dandelions are perfectly capable of reproducing sexually just like the rest of us but, though some do, they seem to prefer the asexual method. Perhaps it's because dandelion seeds are so well-travelled and maybe, at one time, just too many little dandelion parachutes were landing in places where there were no other dandelion mates nearby. Some say it might have been a way for the plant to recover more easily from the ice age. Faced with a life of solitude, it somehow evolved the ability to produce clonal seeds. When you see a group of dandelions together, most of them are clones of the same plant. Just in case these clones turn out to be defective or unable to cope with some kind of change in their surroundings, dandelions have retained the ability to reproduce sexually.

Dandelions are not the only species of plants that love themselves in this way. There are thought to be a few hundred species of plants actively engaging in entirely maternal seed production without the need of a daddy, including my nomination for 'the mother of all plants', the wild bramble. It is believed many more are capable of doing so if they feel the need. So now you understand why most plants might be entirely unimpressed by Dolly the Sheep.

Even plants that don't, as a rule, indulge in non-sexual reproduction, can be forced to do so quite easily. Anybody with a basic knowledge of plants can persuade one to create an identical copy. The reason we can clone plants so easily is all to do with something plants are able to hold onto throughout their lives that we and most other animals cannot – undifferentiated cells.

When our stem cells receive the signal to grow a finger, those cells lose the capacity to become anything else but a finger. That finger grows to a determined length and that's it. If we lose the finger, we cannot re-grow it – it's gone forever. Plants, on the other hand, keep the equivalent of stem cells for their whole lives. It's a necessary defence system for something that is unable to move and must suffer having a bite taken out of it on a regular basis. This remarkable ability to regenerate with relative ease allows us to create perfect clones of plants we like. In this way, we can do away with the unpredictable nature of sexual reproduction. If a plant produces flowers we admire, we can make sure the flower quality isn't altered at all when we make new plants.

We know that as the SAMs reach up towards the sun, they stop every now and then to deposit cells. These cells are indeterminate. They are capable of producing leaves, side shoots, flowers or even roots. Each side shoot, though not dominant over the whole plant, has its own SAM which dominates that particular branch, so there will be lots of little bundles of cells distributed throughout the plant, all undecided about what they will become. Because these cells have yet to decide what they're going to be, we can encourage them to produce both roots and shoots, creating a new plant altogether. Layering is one way of doing this while the mother plant is still attached, but we can do it without the mother plant's assistance.

Auxins are plant hormones, responsible for regulating growth in plants. The same hormones that inhibit the growth of branches lower down when there is a dominant SAM can also encourage roots to form when needed. Under normal circumstances, a plant wouldn't need to form roots along the stem because this would be a very inefficient way to collect water. Therefore, the signal to turn on the development of roots is usually switched off. However, if a SAM finds itself cut off from the rest of the plant, it moves quickly into self-survival mode. The auxins travel down the stem and attempt to form new roots.

For the most part, plants don't need our help for normal day to day activities, provided they are growing in their natural habitat. There are even a few plants that can produce roots from broken stem tips in the wild. However, the odds of a separated stem finding itself in conditions just right for new root generation are not very high at all. The broken tip must find itself the right way up, preferably with the end buried in soil, or at least in contact with water, and the break must be clean enough so disease cannot enter easily. On top of that, it must find these conditions quickly before so much water is lost there can be no recovery. It's no surprise then the vast majority of plants need our help to form roots from cuttings.

## How to clone plants by stem cuttings

The most popular types of cuttings are softwood, semi-ripe wood and hardwood and there's no great mystery as to which type you should try. It's all to do with **when** you take the cutting.

Softwood cuttings are taken from soft, fresh, non-flowering shoots. Plants such as *Coleus, Osteospermum* and *Pelargoniums* are popular candidates for this type of cutting, as is *Buddleia, Hydrangea, Euonymus* and many more. For plants growing outdoors, the time to take softwood cuttings is after the plant has produced soft new growth and before the stems begin to harden – usually around June. You can also take softwood cuttings from over-wintering tender perennials such as *Pelargoniums* after they have finished flowering and you have brought them indoors. Under the right conditions, softwood cuttings root really fast.

Semi-ripe wood cuttings are taken when new stems are just beginning to harden. The wood is reasonably firm and the leaves full. These cuttings are taken from midsummer until early autumn and take a little longer than softwood cuttings to root.

Hardwood cuttings are taken in winter when the plant is dormant. These cuttings take the longest. There are often no leaves on the plant and the cuttings can be left outside.

Whatever type of cutting you make, the method is basically the same for all types:

**Step one:** Prepare your rooting medium in advance. This should be good sterile compost, mixed with something like perlite to encourage it to be free draining. If in doubt, buy pre-prepared compost suitable for rooting cuttings. The pots into which you insert the cuttings should also be sterile to reduce the risk of fungal infections.

**Step two:** It goes without saying, but you need to choose a healthy parent plant which has been well-watered beforehand. Take your cuttings in the morning if you can because the plant generally contains more moisture at this time. Aim to take a number of cuttings as they won't all be successful.

**Step three:** The size of the cutting you take very much depends on the type of plant.

It will be determined by the length of the internodes - the spaces between the nodes. Locate the tip of the stem and go back between 5 -10cm, then make a nice, clean cut. Cutting above a node means you don't leave an exposed internode on the donor plant and avoids die-back on the part of the stem left behind. Later, you will trim the cutting again to just below a node of your choice. At this stage, your cutting is in danger of losing too much moisture, so put it straight into a plastic bag.

**Step four:** Take your cuttings out of the plastic bag one at a time. For softwood and semi-ripe wood cuttings, cut just below a node to a size that will stand up well and support itself when inserted into your rooting medium. Carefully remove all but the top set of leaves. Cuttings taken at a time of year when the leaves are still attached will need some leaves to continue photosynthesis, but too many leaves on a cutting results in critical moisture loss before it's had a chance to grow roots. For plants with large leaves, you can even cut the top set of leaves in half. If the tips of your semi-ripe cuttings are overly soft, remove the tip and retain the next set of leaves down. For winter hardwood cuttings of deciduous plants, you are going to remove the tip, cutting above a node at the top of the cutting and below a node at the bottom. Your cutting should be pencil-thick. Hardwood cuttings are cut with a sloping cut at the top and a straight cut at the bottom. That way, after you've prepared a few, you know which way up the cuttings should be inserted.

**Step five:** If you want to, you can dip the bottom of the cutting in hormone rooting powder. In theory, this powder artificially recreates the hormones that encourage roots to form and should make the cutting root faster but some say it has no effect whatsoever, so it's up to you. Most rooting powders also have the benefit of an added fungicide to help prevent against your cutting rotting away before it's had a chance to produce roots. I personally never use it and I seem to get reasonable success. If you don't have it to hand, you shouldn't worry too much.

**Step six:** Now prepare your rooting container. Hardwood cuttings are going to be left outside so they are best in a large, fairly shallow container so the rooting medium retains moisture when it rains. You can also put hardwood cuttings directly into a trench in a shady corner. In both cases, put a layer of gravel or sand at the bottom of your container or trench so that the end of the cutting sits on it before backfilling with your rooting medium or soil.

Soft and semi-ripe wood cuttings generally go into fairly small pots. Make a small hole with your finger or a dibber near the edge of your pot, then place the cutting into the potting medium and firm around the stem well. Your cutting should be at a suitable depth so the node is completely covered and the cutting can stay upright. Make sure you don't allow any leaves to make contact with the growing medium as this will encourage them to rot. I would normally put a layer of gravel or perlite on top of the potting mix to separate the leaves from the potting medium. Place the rest of your cuttings in a circle around the pot, far enough apart so that none of the leaves are touching.

**Step seven:** Now you have to help your cuttings out a lot. They don't have a root system yet so they're relying on you to provide a steady supply of moisture. Too much water and you risk drowning the cutting or creating an atmosphere where damaging fungi can grow. For softwood and semi-ripe wood cuttings, a nice warm potting medium is helpful so keep them inside and provide gentle bottom heat if you can.

Put your cuttings where they have access to light, but not in direct sunlight since this might dry them out too quickly. Keep the rooting medium moist and spray the leaves with water from time to time. You can cover the cuttings in clear plastic or place them in a propagator with a clear plastic lid. This will retain moisture, but remember the leaves still need air so open up the plastic bag or take the lid off the propagator for ten minutes each day.

**Step eight:**   There's no hard and fast rule about how long a cutting should take to root. In general, it's ten or more days for softwood cuttings; a good few weeks or months for semi-ripe wood; and several months to a year for hardwood cuttings. If you suspect your cutting has rooted, give it a gentle tug and, if there's any resistance, congratulations! As the roots develop, move your cutting to a bigger container to grow on for a while before planting it outside.

For difficult to root semi-ripe wood cuttings, you could try the heel or mallet method. Heel cuttings are taken by carefully ripping away a short side shoot so you take a slice of the harder wood with you. This is often done with conifers. Mallet cuttings also involve taking a small piece of the stem.

# Part Four: Worshipping the Sun

# Reaching out to Ra

If plants were religious, their god would definitely be the sun god Ra. Sunlight means everything to a plant, no plant can survive without it and a plant's ability to manufacture its own food through the process of photosynthesis is the basis for virtually all life on earth. As far as we know, only plants, algae and some types of bacteria are autotrophs - capable of manufacturing their own energy. Every other living thing must consume autotrophs to obtain the energy they need.

The sun has a starring role in the photosynthetic process and sunlight, to a plant, is like the power source that drives a factory – without it, plants cannot make or produce anything. For most plants, photosynthesis takes place mainly in the leaves and this is where the sugar-making factory is at its most efficient. The leaves are like little solar panels, able to capture energy from the sun and use it to power the factory. Once the sugars – glucose and fructose - have been created, the energy captured within the sugars can either be used straight away or stored for use at a later time.

Like any factory system, plants need a

conveyor belt to move their products around and the conveyor belt they use is the phloem which carries sugars through all parts of the plant. It is made up of living cells forming a system of stacked sieve tubes capable of transporting sugars in either direction, extending all the way from the tips of the leaves into the roots. Glucose and fructose are converted into more stable sucrose for easy movement around the plant.

Because sucrose is easily dissolved, the phloem is able to carry sucrose in its dissolved state quite efficiently, rather as you might transport water through a pipe. However, for storage purposes, the sugar molecules need to be bound together into starch. Imagine you have manufactured a load of marbles in your factory and now you need to store them in a warehouse. It wouldn't make sense to have them rolling around the warehouse floor so it's much easier to box them up and contain them until you need them. When plants store up sugars in the form of starch, they are simply boxing them up for convenience and stability. The starch is non-soluble, takes up less space than the dissolved sugars and is less likely to leach out of the plant.

Plants store starch in stems, roots, buds, in tubers such as potatoes, in bulbs and in seeds and fruits. The starches are used for both short and long term storage of energy. Storage in the seeds and fruits provides the next generations with energy when they need it. Storage in the stems represents a convenient location for easy access. Storage in roots, tubers and bulbs means, if the plant is damaged above ground or when the green parts die back in winter, the plant has a protected underground warehouse.

When it is time for the plant to call upon its reserves, the starch is turned back into easily transportable sugars, the phloem conveyor belt switches direction and carries the sugars back out of storage to where they are needed most. The process of moving sugars around in this way is known as translocation.

Of course, the production of sugars would not be possible without photosynthesis and photosynthesis cannot take place without light. It wouldn't surprise you then to discover that plants have the ability to sense light levels and they spend their whole lives reacting to this information. When a seedling first encounters light, it analyses both the amount of light falling on its leaves and the direction of that light. From this information it knows if it is in the presence of other plant competitors or anything else such as a wall or structure which may impede its ability to collect energy from the sun. Armed with this knowledge, it takes steps to ensure it positions itself as close to the sun as possible.

Some plants get involved in a race with the other plants around them to reach the light first. They produce speedier vertical growth and may be taller than usual. In the race to grow faster, their stems become longer and thinner and SAM has less time to stop and deposit bundles of cells at the nodes. The nodes, lateral branches and bundles of leaves are therefore further apart than normal. When this happens, the plant is less sturdy. It may have reached the light in double quick time, but it won't be nearly as strong as a plant that has had to cope with less competition for sunlight.

For other plants, the quickest route to good sunlight might be to move sideways rather than upwards. The same auxins that allow the shoot tips to stretch upwards whilst inhibiting growth elsewhere, can be redistributed to favour one side of the stems. The auxin IAA (Indole Acetic Acid) encourages cells on the shaded side of the stem to elongate which, in turn, allows the plant to bend towards the light. Where there is a moving pattern of light, as the sun moves across the sky, the plant bends and sways from east to west through the course of each day to follow the sun.

The redistribution of auxins is controlled by the shoot tips. Quite how it is triggered is not fully understood. Now, a plant can no more make a conscious decision to produce or move around hormones, than a big-brained, self-aware individual like me. If I want to lift something heavy, or even grow myself a beard, I can't just manufacture some more testosterone or move what little I have around my body. What we do know is this: Charles Darwin put covers over the tips of some plant stems and left them off the tips of others. Those plants without the covers bent towards the light as usual. Those plants with the covers did not bend at all. In other words, if you blindfold the SAMs, you blindfold the entire plant.

Just as we might stop, or move forward very gingerly when we find ourselves in darkness, some plants in the shadow of others do exactly the same. Without enough light to achieve maximum growth, and their SAMs all but blindfolded, they hang back until a chance occurrence, such as the death of a neighbouring plant, allows them to find the sun again. A tree falling in a forest can signal growth spurts in many of the plants occupying the space underneath its canopy.

If we could live our lives at a plant's pace for just a short while, we would be surprised by all the movement we see around us. Plants worship their god so much, they never take their eyes off it, following it around from dawn until dusk.

In extreme circumstances, when a plant cannot see enough of the sun, it slows down the production of chloroplasts. Chloroplasts are the sub-cells within plants which are responsible for the process of photosynthesis. Each chloroplast contains a substance called chlorophyll. If you study light passing through a prism, you will see light is not one single colour, but rather a collection of colours – the light spectrum. Chlorophyll absorbs light energy, but favours light in the red and blue spectrum, so light in the green spectrum is reflected back. This is what we see. Green leaves are not green as such, they are merely reflecting back light on the green spectrum.

Without sufficient sunlight, production of chlorophyll is inhibited. The leaves therefore appear light green or even yellow. With far less energy than others, these pale and sickly looking plants limp slowly towards the light, producing few leaves and little branches until they eventually find the sun again. Once they manage to catch sight of the sun, their god shines on them again and they green up. The stems need time to thicken up and recover, but this should happen with relative ease provided they remain in the favour of Ra.

Young plants have the best chance of leading a long and happy life if they have access to light as soon as possible. That way, there is no check to their growth and their stems are thick and strong from the start.

## Marching bananas

Banana plants are true sun worshippers and

legend has it, they walk towards the sun. As with all legends, there is an element of truth to the idea that banana plants walk. If you mark the location of a banana plant one year and return to it the next, you will find it has moved from its original location. This is because the banana plant is not actually a tree as some might think but an herbaceous plant. In the case of bananas, the vegetation above ground dies down after flowering but the plant is able to re-grow from below-ground stems. A fresh shoot emerges a distance away from the original stem and will usually have travelled in the direction of the sun. This makes it appear as though the banana is walking towards the sun.

In some banana plantations, the plants are laid out in a line running from east to west so they all move south together towards the sun, whilst remaining in a straight line. And there we have it – marching bananas!

# Fixing problems associated with lack of light

We know plants demonstrate short-term reactions to sunlight – or rather the lack of it. When we recognise any of these symptoms, we can take steps to improve things before it's too late.

When a plant has to lengthen its stem to reach better light, the plant's ability to hold itself upright is compromised. It becomes vulnerable to strong winds, heavy rain and even to water from a watering can. For seedlings, it's important they develop a nice strong stem from the start since this will stand them in good stead for the rest of their lives.

If you are bringing on seedlings on a windowsill, you might find they get long and leggy and even bend towards the window. You can turn them each day to expose all sides to equal light, or you can introduce artificial light by training a couple of light bulbs onto the darker side. Putting a mirror or aluminium foil behind seedlings on a windowsill helps reflect the natural light onto the shaded parts.

If it's too late and they've already gone leggy, simply transplant them to a larger pot and plant them deeper – up to the first set of leaves, then ensure they get sufficient light. This will give them a chance to beef up before they go outdoors. Transplanting some types of plants deeper might be risky because the stem could rot below soil level but leggy seedlings are really no use to you so it's worthwhile giving it a try.

Remember to turn houseplants to keep the growth nice and even. A mirror behind houseplants can look really effective. Not only will it reflect the light but your plant looks twice as big.

Outside, you may need to move some plants away from others competing for their share of the light. Consider thinning out tree and shrub branches to open out the canopy. If your garden is very shady, paint walls and fences white to reflect sunlight.

Don't use the mirror technique outside unless you put trellis or something in front of it. Mirrors can be really effective, fooling the eye into believing your garden stretches further than it actually does. However, birds flying towards the mirror are also fooled and can attempt to fly through it. Depending on the speed of its flight, this may knock out a bird or even kill it.

## Blanching vegetables

Some vegetables taste sweeter and are less coarse if you encourage them to stop producing chloroplasts. Vegetables such as celery, leeks and endives all taste better when the stem and leaves are pale. To achieve this, you simply block out the light at the appropriate time.

For leeks, it's relatively easy. Leeks should be started from seed in a separate seed bed. When the seedlings are around 15 to 20 cm tall, prepare the permanent planting site by making holes deep enough to take all but the very tip of your entire seedlings and not just the roots. If you use a broom handle to make the holes, you've got the right width as well. Drop the leek seedlings into the holes and, without backfilling the holes, water them in so the water encourages the soil to settle loosely around the seedling. Your leeks will grow as thick as the hole and nice and white for the first 15cm because the light has been excluded.

For celery and endives, wait until you are about two to three weeks away from harvesting. At this time, you need to exclude light from the lower leaves. You can either tie up the leaves so you get white stems in the middle of the plant or you can slip a pipe over them, or wrap them in newspaper. Excluding light from celery and endives makes them less bitter – at least to taste anyway. The plants themselves will be feeling bitter as hell as it's not in their nature to grow this way.

# Using a Ladder

In the competition to achieve maximum sunlight, trees are the obvious leaders. With their strong, woody stems, they stand head and shoulders above most plants and can always be guaranteed good access to sunlight. But it takes many years and a great deal of energy to produce a stem strong enough to stand up to the elements without bending. Some plants have evolved the ability to climb into the light without investing huge amounts of energy in producing a thick supporting stem. Instead, they use the support of other plants or structures to get up to that much-needed sunlight.

The climbing plants are not necessarily shade-tolerant, but many will be born in the shade of other plants. The seedlings of some climbers behave in the opposite way to most emerging youngsters. They do not search for light in the first instance but, instead, look for shade. By moving towards shade, they increase the chances of encountering something useful to use as a ladder. Most climbers like to have their feet in the shade and their heads in the sun.

Twining climbers grow in a motion

describing a circle or ellipse in the air. A 'twiner' indulges in this circular movement even before it has found a support. Like a spinning rope it spins slowly through the air until it encounters something just the right thickness to wrap itself around. Some twining climbers can wind around thick tree trunks, while others prefer thinner stems. Though most climbing plants rarely directly damage a host plant, the twiners in particular, can grow so heavy they pull down branches and sometimes the entire plant they are climbing up. This, as you might imagine, is just as inconvenient for the climber as it is for its host plant so it's a lucky climber that finds the perfect host. On a positive note, it is possible for a twiner to actually support the host plant by acting much like a guy rope, providing additional protection for the host plant during strong winds.

Other climbers produce modified leaves or stems (known as tendrils) to enable them to grab onto and curl around a support in the same way a mountain climber utilises handholds to assist with his ascent. The tendrils are highly sensitive to touch so, as soon as they encounter any resistance to their free movement through the air, they begin to curl just as we might curl our fingers around a rope or a ladder. By coiling up the tendril nice and tight, the plant is able to pull itself closer to its support. *Clematis* falls into this category, as do sweet peas and the slightly tender climber, Passionflower *(Passiflora caerulea)*.

Virginia creepers are like Spiderman. They, and others, form forked tendrils with adhesive pads at the end which can hold onto walls and the trunks of trees. Those climbing plants adorning the walls of older houses are generally of the types which climb by means of an adhesive pad.

Most ivies produce little aerial roots on their stems which cling tightly to a surface. How they do this is actually quite complicated and really very clever. When ivy establishes it has made contact with a suitable surface, it looks for any gaps in the surface, big enough for the aerial roots to occupy. Once they have entered a space, the aerial roots grow root hairs which then thicken and act much like the wall fixings we use for shelves. As the root hairs expand, they fill the space, creating a tight fitting. On top of that, the ivy secretes a glue-like substance which dries hard, further increasing its hold. When the root hairs die, they remain surrounded by the glue, behaving pretty much like a wall plug. This plant is very serious about holding onto its support and it's no wonder ivy is so difficult to get rid of.

Other plants don't so much climb as scramble towards the sun. They have barbs or thorns which act as grappling hooks to drag them ever upwards. The thorns don't just help them reach greater heights, they help them hold onto their support, though these thorns and hooks will not be as effective at holding on as other methods of climbing. Rambling roses fall into this category.

Whichever way a plant climbs, it is easy to see how the ability to do so can have major benefits to a plant born in the shade and some can reach a position of advantage remarkably quickly. Climbing plants don't belong to one particular group of plants either. Rather, they seem to have evolved within a number of plant families in response to certain environments. It's no surprise then that a large number of species of climbing plants are found in the tropical rainforests where the ability to ascend to the canopy above has obvious advantages. As Tarzan so expertly demonstrated, these vines can be extremely useful to other animals as well, forming bridges between trees and allowing animals to traverse the canopy without having to go to ground.

# The vine that can produce furniture and bones

The plant with the record for the longest stem in the world is a spiny climbing palm. The stem of *Calamus manana* can reach as long as 200 metres and is one of a number of species that gives us rattan used to make furniture and baskets. In

Faenza near Bologna, there's a herd of sheep walking around with bones made with chemically-treated rattan wood implanted in their legs. Unlike other types of bone implants, it would appear as if the porosity and structure of rattan is allowing the natural bones from the sheep to merge with the new bone made from rattan.

If all goes well with the sheep, it may be possible to implant these 'wooden bones' into humans and, with the incredible strength of rattan wood, the manufactured bones will be capable of supporting a human body without having to be replaced over time.

# The vampire vine that smells its victims

There are between 100 and 170 species of dodder plants in the world and all have a grim reputation. The red or yellow dodder is born without chlorophyll and doesn't have the ability to photosynthesise. Unable to harvest sunlight, for its whole life it must suck energy from other plants. With only a small amount of energy stored in its embryo, a newly-emerged dodder needs to find a suitable host within a week or it will run out of stored energy and die, but the dodder has hosts it prefers over others so it can't simply head for shade and hope for the best. Instead, the dodder smells its victims. While we humans can recognise the distinctive smell of some plants such as tomatoes, the dodder has a keen sense of smell and can smell plants we can't, or rather it has an attraction to the chemical signals given off by certain volatile organic compounds.

Once settled on a host, dodder twines itself around the plant. It has little bumps on its stems capable of penetrating host plants and sucking energy and nutrients directly from them. When it has sunk its fangs into a suitable host and, since it will now live entirely off its host, it has no further need of roots and they wither away. Since it cannot photosynthesise, it has no need of leaves either.

Dodder can move from one plant to another so it's no matter to dodder if it kills its host providing its twining stems are able to locate another in time.

## Does Ivy Damage a Wall?

It has often been said that ivy damages the walls of houses by breaking up the mortar with its aerial roots. This isn't strictly true for houses where the walls are sound to begin with. If the mortar is in good shape, ivy is unlikely to be able to find any holes or cracks it can exploit. Mortar mixes using Portland cement, as opposed to lime mortar, should be strong enough to support ivy without any real concerns so, if your house was built post 1930, it should be fine. The real reason ivy gets its bad reputation is because it is quick to make use of any weaknesses in a wall and, once it's covering a wall it's hard to spot any structural defects until it is too late. A wall covered in ivy is also harder to maintain, though some say the leaves of ivy may give some protection from the weather.

If you feel you have to remove ivy from a wall, remember the root hairs initially responsible for the ivy's hold on the wall are already dead and the plant is being held in place by the glue it produced around the root hairs. So the often-told advice of severing the ivy at ground level and waiting for it to 'let go' really doesn't work. Deal with ivy exactly how you would deal with the removal of a shelf from a wall fixed using wall plugs. Either gently cut the ivy away and leave the plugs intact or rip it out and be prepared to repair the holes left behind in the mortar. Any aerial roots remaining glued to a wall after you have removed ivy can only really be removed by scrubbing vigorously with a wire brush.

Boston ivy and Virginia creeper don't have destructive aerial roots and their sticky pads cause less damage to walls. They can be removed from a wall by severing the main stem at ground level and waiting until the leaves have turned brown. After that, it should be quite easy to pull the plant away from the wall.

# Using climbers in the garden

Most climbers are sold container grown and you can pretty much tell by looking at them in the pot how they climb. As a general rule, those with adhesive pads or aerial roots like to be against a solid surface. Most woody twiners need something relatively strong to wrap around. Those plants with tendrils will want something a bit thinner to get a hold of and the scrambling types may need to be tied in.

**Over pergolas and arches:**

For pergolas and arches with trellising on the sides, you can use any kind of climber. If using a twiner, stick to the more delicate ones so your trelliswork doesn't get destroyed as the stems thicken. *Clematis* has leaf tendrils and grabs onto trellising nicely. When it reaches the top it travels horizontally across the roof so it's a great choice for an arbour. Any climber using tendrils will do fine over a pergola but you may need to provide some wire or surround the posts in netting if you don't have trellis on the sides. A twiner like *Wisteria* is best used on very sturdy arches built from open posts as *Wisteria* has been known to bring down porches and pergolas and will probably break trellis. If you want to use a scrambler such as rambling rose, then you'll need to tie it in as it grows since the smooth edges of pergolas and arches aren't really enough to give a scrambling climber a decent thorn-hold.

**Against walls:**

Climbers with aerial roots or sticky pads are a good choice for walls. *Parthenocissus quinquefolia* (Virginia creeper) and *Parthenocissus tricuspidata* (Boston ivy) are good choices for autumn colour and *Hydrangea petiolaris* (Climbing hydrangea) is a favourite of mine for clothing garden walls. It's got aerial roots so it needs something strong to hold onto. Wall shrubs aren't true climbers, but have a preference for leaning against a wall or other structure. If you want to control the direction they take, you need to tie them into wires and pegs set into the wall. I recommend the stunning, fiery *Pyracantha coccinea*. Train the side branches out onto wire rows underneath a window and dare burglars to try and climb in over its lethal thorns.

**To hide something such as a tree stump:**

Definitely scrambling climbers are best here. They'll cover a tree stump in no time. Golden hop (*Humulus lupulus)* is a good choice as long as you keep it under control.

**On an obelisk or wigwam:**

Beans are twiners so they like to wrap themselves around a wigwam of bamboo poles. Peas and sweet peas need netting or something woven through your obelisk as they use tendrils for stability. If you try to make peas climb up a bamboo pole without netting, the pea plants usually prefer to grab onto each other, rather than the pole.

**Through trees:**

It is possible to use some climbers to grow through trees and that way you get a double season of interest – when the tree flowers or changes leaf colour in autumn, and when the climber flowers. You need to be a bit careful since, though they don't set out to harm the tree, some climbers may be too vigorous and may compete with your tree for light. I wouldn't use ivy on a tree – it's just too strong for most trees, but some of the tendril types like clematis are okay. As the trunk of the tree will be too thick for the tendrils to curl around, you'll need to provide some initial help to get the climber into the thinner branches. You can do this by attaching string to a peg on the ground and running the string line up the trunk and into the branches. Alternatively, you could use netting, but don't wrap netting all the way around the trunk as you'll prevent the trunk from thickening.

**As ground cover:**

If there is no support available, some climbers are just as happy to run along the ground and they can be very effective growing in between other plants. The obvious one that springs to mind is English ivy (*Hedera*) but *Clematis*, Virginia creeper and nasturtium can do this as well. As a general rule, twiners don't make good ground cover. Delicate climbers with aerial roots, adhesive pads or tendrils are best for growing along the ground as long as they are not allowed to swamp neighbouring plants.

# Born in the Sky

Some plants can't be bothered with all that mountaineering to reach high up into the canopy of trees. After all, why bother to climb up a tree when you can be born at the top in the first place? The epiphytes, or air plants, live their whole lives clinging to the high branches of trees and never even get a look at the ground below. Their seeds are carried into the canopy on the coats, or in the droppings, of birds and climbing mammals. A tiny piece of leaf litter or debris in the crook of a branch is all it takes for the seed to receive enough moisture to germinate.

Most epiphytes exist in humid rainforests where their roots and leaves can absorb moisture from the air. In the cooler climates, we meet these plants in their simplest form - as mosses, liverworts, lichens and, sometimes, as ferns. Their presence in woodlands is said to be an indicator of good air quality. Since these plants are essentially air-living, they are going to need good air in order to thrive.

It is often said, if you want to determine which way is north, you should look at moss growing on trees. Whichever side of the tree the moss is growing on, is facing north. While there may be some truth in this, navigating by moss alone is likely to get you completely lost because moss can and does grow on any side of a tree where it finds the conditions it prefers. Therefore, it may grow on the west-facing side or even on the south-facing side provided there is adequate shade, sufficient moisture and protection from the desiccating effects of the wind. While we might assume all these conditions to be met on the north-facing side, it isn't always the case so bring a compass with you just to be on the safe side.

Most epiphytes do no harm to the plants they are living on and most are largely beneficial to other insects. However, mistletoe, though not a true epiphyte but epiphytic in habit, is a parasite that relies not only on photosynthesis for its energy requirements, but also sucks some of the energy from its host plant as well. It is quite possible for a heavy infestation of mistletoe to kill the host plant but mistletoe isn't all bad. It is especially beneficial to birds because they are able to eat the poisonous berries without any ill effect and they can use mistletoe as a shelf on which to build their nests.

Some flowering epiphytes are incredibly beautiful and the ones we encounter most are those that have been taken from their rainforest habitats and forced to live in our homes. These lovely houseplants are bromeliads and orchids.

Most epiphytic bromeliads have leaves which are formed into funnels - very useful for collecting and storing water. The water collects in little pools at the base of the leaves and, in the wild, these pools are often inhabited by crabs, salamanders and tree frogs. These animals use bromeliads as babysitters, to protect their young and keep them safe from predators. It probably helps out the bromeliads too since the little ones deposit waste in the water which, in turn, feeds the plants they live in.

Bromeliads usually produce quite insignificant flowers, favouring instead to apply make up to specially adapted leaves (bracts) surrounding the flower and some of these bracts can be spectacularly colourful which is why we like them so much as houseplants. If you're going to live so high up, you'll need to try hard to attract attention so the bromeliads go to great lengths to do so. Bromeliads come with all kinds of colourful bracts from bright yellow, through to vibrant pink.

Epiphytic orchids have specially adapted roots that act like sponges to absorb available water. They are capable of absorbing and storing even the tiniest drop of water. These orchids can produce an array of weird and wonderful flowers designed to attract all manner of flying insects.

# Ants in my plants

Living a life high up in the branches of trees may be a good way to ensure regular access to sunlight, but access to nutrition is an entirely different story. The epiphytic ants-nest-plants have formed a relationship with ants benefitting both the plants and the ants to such an extent that neither does quite so well without the other.

To begin with, the ants-nest-plants build a home intended to be the ultimate in ant residences. They produce a tuber (a swollen stem) with chambers and tunnels inside which is just the kind of dream home a colony of ants might be looking for. If ants set about excavating a home within a plant for themselves, they would want it to look just like the home the ants-nest-plant provides for them. However, the ants have no influence over the architecture since ants-nest-plants produce this structure with or without the presence of ants. Rather the plants build the homes and then hope they will be lucky enough for a colony of ants to want to live there.

In some ants-nest-plants, the chambers forming the rooms of these ant homes even come already furnished by their landlords – the nurseries with smooth walls and shelves on which to store pupae and the toilets with warty structures on the walls. In the nursery chambers, the ants tend to their children. In the toilets, they deposit their waste and that's just what the plants want because the warty walls in the toilet block are able to absorb nutrients. So the ants get a home and their rental is paid to the plant in the form of nutrients.

Provision of nutrients is not the only way the ants satisfy their landlord. Having made their homes in the plant, they defend it as vigorously as they would any home and so the plant is protected from leaf-eating predators. In return for their defence of its leaves, the plant does what it can to protect the ants by producing leaves modified into spines thus allowing the ants to wander about in relative safety behind a barbed wire fence. At flowering time, the plant often rewards its tenants with a little sugary drink in the form of nectar.

In a twist in the tale that resembles the plot of Medicine Man, there is an interest in some species of ants-nest-plants as a potential cure for cancer. Local people have been boiling the tubers to use as cures for a variety of ailments for centuries. It would seem that a derivative from one of these plant species was able to induce the cell death of breast cancer cells and another was seen to discourage the growth of certain bacteria.

## Caring for epiphytes

If you buy an orchid for a houseplant, you often find it's in a transparent pot. When you realise that some orchids are epiphytes, then the reason for the transparent pot is pretty evident. These plants aren't used to having their roots kept in the dark since they spend all their time on trees without the benefit of deep, rich soil. If you have your roots exposed to light like this, it would make sense if the roots as well as the leaves could carry out photosynthetic duties and it is widely accepted the roots of epiphytic orchids do just that. So, if the roots of your orchid are green, you don't have to worry too much since this is just evidence of photosynthesis.

There is another advantage to having a clear pot. You can see if the roots are becoming unhealthy or when you need to water. Epiphytic orchids should be allowed to dry out between watering as this reflects their growing conditions in the wild. If you look through your clear plastic pot and see the roots are not green or white, but brown and mushy, this is a sure sign of over watering.

Over time, the roots begin to emerge over the top of the pot and you can replant into a bigger transparent pot, but the growing medium in the new pot should be extremely light and airy because it's not a plant used to growing in heavy soil. When your orchid has finished flowering, cut the flower stem back to a node below the original flower and wait a while. Eventually, a new flower will emerge.

Epiphytic bromeliads form a rosette of

leaves around a central cup designed to hold water and should be watered into the funnel formed by their leaves as well as occasional watering of the soil in the pot. The plant prefers to have water in its funnel at all times but, if the water has become stagnant and smelly, you should empty it out with a syringe and replace it with fresh water. Allow the soil to dry out between watering, but always keep some water in the funnel.

Bromeliads flower just once and the flowers are usually to be found as insignificant clusters inside the cup. It is the colourful bracts surrounding the flower we are most interested in. After flowering, the bracts remain on the plant for several months, during which time the plant produces new cloned plants at its base known as 'pups'. When these pups have developed a recognisable cup, you can sever them from the mother and re-pot them in free-draining compost. Don't be too upset if, after producing her pups, the mother plant dies. This is the natural order of things and, of course she's not really dead since she lives on in her clones.

If you are able to collect rain water, all epiphytic houseplants prefer to be watered with this, rather than tap water. This is because rainwater contains a small amount of nutrients and the epiphytes in particular are expert at extracting these trace elements.

Finally, if you can, take your epiphyte with you when you go for a shower. In the wild, tropical epiphytes inhabit a foggy environment, wrapped up in a warm mist. Placing them in the bathroom while you have your shower is a really nice treat for them and reminds them of home. If you can provide them with enough light, a bathroom is the best place to keep tropical epiphytes permanently only careful it doesn't get too cold in there.

# Living in the Shadows

Without sufficient sunlight, plants become weak and may even die. Many go to extreme lengths to alter their normal growth pattern so they can reach the sun, but what about plants growing in the shade? In the shadowy understorey of woodlands and forests everywhere, there are plants growing quite happily without any of the symptoms we associate with lack of light. These are the shade-tolerant plants and they come from all kinds of plant families and in all shapes and sizes. They survive because they have adaptations which allow them to live in low levels of light.

All plants have a light compensation point. Simply put, this is the point at which the amount of energy gained through photosynthesis is the same as the amount of energy and carbon dioxide lost or consumed by the plant in order for it to grow. The human equivalent of a light compensation point would be our calorie requirements. The average, relatively inactive woman needs between 1600 and 2000 calories to maintain her weight. Any less and she will lose weight, any more and she will gain it.

While it may not be desirable for the

average woman to gain weight, plants are perfectly happy to do just that. Green plants prefer a net amount of light higher than their light compensation points. A plant growing at light levels below its light compensation point may have to sacrifice an energy-sapping function such as flowering. That's why you get fewer flowers on sun-loving plants when you put them in shady gardens.

Shade tolerant plants have lower light compensation points than sun plants. In other words, they need fewer calories and they have the ability to photosynthesise more efficiently than their shade-hating counterparts. Many shade-tolerant plants have larger leaves. If you think of a leaf as a solar panel, then it follows that the bigger the solar panel, the more efficient it will be at capturing light. Inside the leaves of shade plants, the chloroplasts crowd near the surface in an effort to grab more light and this can result in very dark green leaves. When you put a dark green shade plant in the sun, it often turns yellow to let you know how unhappy it is.

There are some species of shade-tolerant plants with red or yellow leaves. This is not because they have less chlorophyll, it's just because the chlorophyll is hiding behind other pigments. Yellow carotenoids are found in the chloroplasts alongside the chlorophyll and they assist with photosynthesis by absorbing blue light thus reflecting back yellow light. Red anthocyanins occur in all parts of the plant and absorb green light, reflecting back red. Green light is least favourable to a plant but, if you're growing in the shade, you have to take what you can get. Anthocyanins may play a part in protecting the plant from harmful UV rays. They may also make the plant appear unpalatable to grazing animals. Shade-tolerant begonias have red on the undersides of the leaves to bounce back any light missed by the chloroplasts on top of the leaf. Despite the fact that some shade-tolerant plants may have different coloured leaves, when a red-leaved plant finds itself in deeper shade than it would like, it adapts by producing more chlorophyll and its leaves start to turn greener. This is why some red-leaved plants turn a muddy brown in too much shade as the green colour fights for dominance over the red.

Because of the energy it takes to flower, plants adapted to growing in the shade might have smaller, insignificant or less colourful flowers. Some, like the bromeliads, produce bracts surrounding a fairly insignificant flower. The leaves are able to attract pollinators without the plant having to waste too much energy on the flower. Some shade-lovers may also choose not to flower every year so they have more time to build up the energy they need for flowering. In very low light levels, plants have sufficient light to produce vegetative (leafy) growth but not flowers and this is the main reason why many of our houseplants don't flower. There just isn't sufficient light for them to do so. In their natural habitat – often rainforests – many shade plants will wait until a tree falls, giving them the light boost they need to flower.

There are certainly advantages to having a low light compensation point. The shade-tolerant plants may have to live in the shadow of other plants. They may have to rely on falling crumbs of light which aren't absorbed by the plants higher up, but they are a lot less likely to suffer from loss of moisture and they get good protection from the wind.

## *Aspidistra* – the plant you cannot kill

Victorians loved the *Aspidistra*. With their cold, draughty and dark houses, it was one of the only plants they could rely upon to survive. Native to the understorey of East Asian forests, it can survive extremely low light levels and handle temperatures as low as -2 degrees Celsius. It is pretty tolerant of gas, heat and dust which explains why the Victorians loved it so much since they could even place this plant close to the fireplace.

You don't even have to worry if you forget to water your *Aspidistra* because it's relatively drought tolerant as well. In fact, the *Aspidistra* plant can survive many extremes except strong sunlight which, being a shade plant, *Aspidistra* positively hates. It's no wonder its nickname is the Cast Iron Plant! And you can tell by looking at *Aspidistra* that it's a perfect shade-tolerant plant. The dark green, shiny leaves grow up to 60 centimetres and are well-designed to capture every bit of available light.

Unfortunately for the *Aspidistra*, its legendary ability to survive extreme neglect has led some people to believe it actually thrives on it. This isn't strictly true. It may put up with worse conditions than most but it doesn't particularly like it. Just because it can handle a certain level of cruelty doesn't give us the right to treat it badly. Treat your *Aspidistra* well, water it regularly, keep it out of the sun and stick it outside in the summer rain now and again and it will be with you for years. It'll have to be because there is a trade-off to having such an obliging plant. *Aspidistras* grow very slowly so you won't get a truly magnificent plant for quite some time.

## Gardening in the shade

Having a shady garden is often seen as a disadvantage and it will be if you constantly hanker after the plants you can't grow. Bright, showy flowering plants such as *Agapanthus* – the African lily – will probably give you a flower or two but they won't produce the kind of show they would if they were in full sun. On top of that, you'll find that flowering plants like these won't grow straight and they'll bend away from the shade. I do grow *Agapanthus* in my shady garden because I love it so much, but it's not a happy plant and I always feel a bit sorry for it as I watch it stretch towards the light. If you have a shady garden, you should seek to avoid this kind of plant cruelty and try to grow plants that will appreciate the circumstances.

If your garden is shaded by trees, most of the flowers you can expect are the pre-vernal types, i.e. those flowering in spring or sometimes in autumn. You won't get an explosion of flower colour in summer, but what you can have is a garden full of pretty flowers in spring followed by lush foliage throughout the summer and then some more flowers at the tail-end of autumn.

My shady garden is dominated by a huge common lime tree - *Tilia x europaea* to give it its Sunday name - which obligingly allows me to climb it from time to time to sit in one of its branches, and features on the cover of this book. Without it, my tiny garden would be bathed in light and I could grow all kinds of exotic flowering plants but I would never even contemplate cutting down my majestic lime and so I have come to appreciate the delicate little flowers of spring as well as foliage over flowers.

Choose large-leaved plants for your shady garden and go for as many shades of green as you can find. Plants with big foliage such as *Fatsia*, *Hosta* and the majestic tree fern Hosta andHwould struggle to survive in a sunny site so think yourself lucky if you've got a garden shady enough to accommodate them. You can use some plants with coloured foliage to give you a good spread of colour too, but these coloured leaved varieties are best grown in the lightest part of your garden. Plant breeders have been playing around with the colour of *Heucheras* for a while now and you can get some amazing foliage colours from zesty lime-green to dark purple. Not all *Heucheras* like the shade but you can find enough varieties that do.

If you focus on leaf shape as well as foliage colour you'll end up with a really fascinating garden. Leaves can be linear – long and thin; palmate – like the palm of your hand; cordate – like a heart; saggitate – like an arrow head; spatulate – like a spoon. They can be oval, oblong, round or shaped a little like a diamond. If you mix leaf colour, size and shape, you will always find something interesting to look at.

In the vegetable plot you should generally avoid any plants grown for fruit as they may need more sun energy than you can give them. However, you can successfully grow many of the leafy vegetables like spinach and kale and, with just a few hours of sun a day, you can also grow broccoli, cauliflower, peas and beans. There are certain advantages to growing vegetables in the shade including the fact you won't have to water as often. Vegetables that have a tendency to 'bolt' – to flower when you don't want them to - are less likely to do so in the shade.

# Too Much of a Good Thing

Plants crave sunlight but, like us, they can be damaged if they get too much sun. When sunlight hits the leaves of a plant, it excites the electrons in the chlorophyll creating the energy required to complete the process of photosynthesis. However, too much excitement can cause the cells of the plant to break down so plants must be careful to avoid an overdose of strong sunlight. Many plants have developed some clever adaptations to allow them to live in areas with little or no shade.

The first defence against the ravages of the sun is to reduce the leaf surface. Sun-loving plants will, of course, develop smaller leaves – the smaller the solar panel, the less chance there is of damage. The leaves of sun-lovers may also have an altered angle. A shade plant faces up to the sun, exposing all of its solar panels, whereas a plant in full sun might change the angle so the leaves are held vertically or part-vertically. Some trees in the rainforests hold their lower leaves horizontally when in the shade, but hold the leaves vertically in the fully-lit canopy. Plants can change the angle of their leaves in response to the intensity of the sun when, for example, they are faced with a bright shard of temporary light as the sun moves across the canopy and finds a space. This reaction is akin to the way we shield our eyes when the sun moves out from behind a cloud.

High up in the mountains, the alpine plants have to contend with conditions that give them no shade from other plants, as well as the drying effects of the wind. Their tiny leaves reduce moisture loss and prevent them from absorbing more sun than they can handle. They stay low to the ground to avoid strong winds as well.

Cactus species have leaves modified into spines which do not photosynthesise at all. Instead, the spines surround a fleshy, moisture-storing stem. The green stem can then photosynthesise under the protection and shade of the spines. The more spines a cactus has, the better it will be protected from the sun.

Many sun-loving plants have fine, silver hairs on their leaves. Like the spines on a cactus, these fine hairs help to shade the leaves and reflect the sun. Any garden plants with a silvery blue appearance come complete with their own sunscreen and will expect to be placed in full sun.

Without these adaptations, plants growing in excessive sun would over-photosynthesise. In other words, they would absorb sunlight much faster than they can convert it to sugars and use up too much carbon dioxide. Without carbon dioxide to grab onto, the plants hold onto oxygen instead, causing dangerous levels of oxygen to build up and killing cells. Plants produce antioxidants which, amongst other things, help them to counteract the harmful effects of the sun. In fact, so do we, but perhaps not quite as efficiently as certain plants. That's why we're often encouraged to eat super-foods like tomatoes, red grapes and cranberries because they are considered to be high in antioxidants. We can rely on these plants to help us top up our own levels of antioxidants.

Most plants photosynthesise using the C3 pathway. In other words, they incorporate carbon dioxide into a compound containing 3 molecules of carbon. C4 plants incorporate carbon dioxide into a 4 molecule compound and store it in specially adapted cells. When they have used up all the available carbon dioxide, they can remove the $CO_2$ from this additional storage space and use it to continue to photosynthesise under conditions other plants would find impossible. These plants are able to avoid the build-up of oxygen that would otherwise kill a normal plant.

While plants may love the sun, they know too much of a good thing can wind up being a bad thing and, just like us, they can even get sunburn. Occasional over-exposure to the sun is an inconvenience but constant exposure without protection can be deadly.

# Sugar cane, an extremely efficient photosynthesiser

As someone who takes three sugars in their coffee, I have to love the sugar cane plant. In terms of photosynthetic efficiency, the sugar cane plant can definitely boast the highest. It is capable of around 8% efficiency in converting sunlight to sugars as opposed to between 1% and 2% for other cultivated plants. One of the reasons it can do this is because it's a C4 plant. C4 plants evolved when they stepped out of the forests millions of years ago and into the strong sunlight. There are around 7,600 species of C4 plants and this isn't all that many when you consider the best estimate of the number of species of plants in the world is 400,000 and counting, so they really are quite rare. Sugar cane and its C4 counterparts are said to have evolved fairly recently as compared to other plant types and this might have been in response to the increasing levels of oxygen and decreasing levels of carbon dioxide that occurred as our atmosphere began to settle down.

In any event, the amazingly high photosynthetic efficiency of sugar cane means just one thing to us – better photosynthesis, more sugar and anyone with a sweet tooth has to be grateful for that. Before the discovery of sugar cane, honey was the only sweetener in the countries beyond which it grew, so sugar cane was quite a revelation for those who had never seen it before.

This highly sought-after plant grows best in tropical and sub-tropical climates and our universal love of sugar makes it a good source of income for the countries in which it grows. France gave up its share in Canada to Britain in order to protect its sugar-cane islands and the Dutch gave up New York for their sugar canes. Brazil is now the world leader in the production of ethanol from sugar cane and has made it mandatory for all drivers of light vehicles in Brazil to use either pure ethanol, or a blend of ethanol and petrol, to power their vehicles.

Sugar cane has a dark history. Wealthy landowners in Caribbean islands not only enslaved the sugar cane plants in massive plantations, they used human slaves to grow and process it. Slaves produced the sugar that was traded for cash to buy more slaves and so the plantations got bigger and bigger. Around 4 million men and women were taken from their homes in Africa and brought to the sugar plantations.

Conditions in the sugar works were particularly terrible for the slaves. The canes had to be fed into crushers by hand by over-worked and exhausted men and women. Often, their hands would get caught in the crushers and they could be pulled into the machines. It is said that there was always a man on standby with a machete, ready to cut off the arm of anyone who got caught in the machines before they were dragged in completely. In the boiling houses, the atmosphere was unbearably hot and burns were common when the hot sugar got splashed onto the skins of the people working in them.

Sugar later became a target of anti-slavery campaigners who organised a boycott of slave-produced sugar, putting a financial squeeze on plantations owners. When slavery was finally abolished, plantation owners took a massive hit and, thankfully, many went bankrupt as a result.

Sadly, conditions for some workers in sugar cane fields are still way below the standards enjoyed by most employees. Investigations into sugar cane production in the Dominican Republic have brought forth claims of starving workers, child labour and people working under armed guard.

## What to plant where

You don't need to look up a list of plants tolerant of shade or a list of plants for full sun. You don't even have to worry if you lose the label that goes with your plant. In general, you can tell just by looking at them which plants you should plant, and in what location.

In the Northern Hemisphere, a south-facing garden will generally get the most sun. It will support all full sun plants and, by full sun plants, I mean those preferring between five and six hours of direct sunlight per day. Since more sunlight means more energy, and it takes a great deal of energy to produce a flower, in a south-facing garden, you can grow a huge amount of different flowering plants. Many plants which like full sun cope perfectly well in gardens without a south-facing aspect as long as they're not placed in permanent shade. In other words, if, as the sun moves through your garden, your plants receive some direct sunlight and a good few hours of indirect light, they'll do just fine. Watch how the sun moves through your garden and place plants with big, showy flowers in areas that receive the most sun for the longest time.

Plants with large, dark green leaves need to have the protection of at least some shade. Too much sunlight and they may get stressed. If a big, leafy plant such as *Fatsia japonica* looks yellowish and sickly, it's probably suffering from too much sun. Give any plants with large leaves the protection of a wall or plant them in shade.

Most climbing plants like their roots to be in the shade and their stems and leaves to be in a good amount of light. This more accurately reproduces the conditions they have adapted to in the wild.

Plants with small leaves or those with grey or silver leaves won't do very well in the shade of a tree. Small leaves and a surface of fine hairs are indications of adaptation to strong sunlight and silver-leaved plants are wearing a white T-shirt – just as we do in sunny weather.

Alpine plants don't expect to be shaded out by a tree, so give them maximum exposure to make them feel right at home. An alpine is fairly easy to recognise without a label or reference book. It has small leaves to reduce exposure to the sun, a shallow root system to cope with the shortage of soil on mountainsides and rock faces and is very low-growing so that it can duck out of the kind of strong wind it can expect to find at high altitudes. Alpines are often grown surrounded by light coloured gravel and you can use gravel around quite a few sun lovers to bounce more sunlight back onto the plant. Sun lovers usually expect to find themselves in relatively dry conditions so using gravel instead of damp soil around the necks of these plants helps them feel a bit more at home.

Variegated leaves are those leaves with two or more colours, often green and white. The white parts of the leaf contain no chlorophyll and plants with variegated leaves rarely occur naturally. They are the result of our exploitation of what would be a disadvantageous mutation in a wild plant. When we see such a mutation, we clone these plants to preserve their markings but, in the wild, it is more than likely that, through sexual reproduction, these plants would lose their accidental variegation and revert back to more efficient green leaves. Because they're not exactly natural, variegated plants are fussier than most when it comes to light. Without a good amount of light, they lose their variegation but too much light and, because the white parts have no pigmentation, they get stressed out. Plant variegated plants in areas where there is good indirect light, neither in full sun nor deep shade.

Plants with red leaves can be a little more difficult to place. Some red-leaved plants thrive in full sun, others don't. As a general rule, plant your red-leaved plant where it will get a good amount of sun, but also a little rest for part of the day. East or west facing or lightly shaded by a shrub that isn't too dense is good for red-leaved plants.

Keeping your plants happy in terms of light levels is easy and it's certainly a time when listening to your plants really comes into its own. A plant will tell you it's not happy for any number of reasons and there may be times when you don't have a choice but to keep your plants under less than favourable conditions. Since sunlight is so important to them, you really owe it to them to do the best you can because plants growing under light levels they don't like will tell you about it pretty quickly.

After you have brought home a new plant, keep an eye on it for its first year and don't be afraid to move it around the garden or, in the case of houseplants, from room to room until you're sure it's happy. You'll be amazed at how well a plant responds when you satisfy its particular light requirements.

If I don't move my *Agapanthus*, it will spend the rest of its life hunched over and unhappy instead of straight-backed and beautiful, so I will have to find space for it in my tiny front garden where it can get more light. My front garden is also where some of my houseplants spend their summer holidays. Just like us, a holiday in the sun can work wonders for houseplants and continues to have benefits long after the holiday is over.

# Part Five: Getting Enough to Drink

# The Role of RAM

It's not enough for a plant to ensure it has sufficient sunlight because photosynthesis can only take place in the leaves if there is water present. For our young plant, RAM (the meristems located near the end of the roots) must get to work quickly to establish its underground network. Just as SAM is responsible for the stem, branches, leaves and flowers, so RAM controls the development of the underground network. Some parts of the roots will be used to store the sugars produced by the leaves so the plant can survive winter. Other parts will anchor the plant to the ground to prevent it from being pushed over but, before all that, the initial emergent root must find water as soon as possible. RAM must travel through the darkness, sniffing out water and mineral nutrients and, once it has found water, it must ensure transportation of the water into the leaves.

RAM contains all the undifferentiated cells essential for root development and later it will allocate certain functions to these cells. Darwin believed it was the location of the plant 'brain' so it is not surprising therefore that the plant wants to protect such an important package and it does so by way of the root cap. The root cap is a group of hardened cells surrounding the tip of the root, protecting it and allowing it to tunnel through the dirt unharmed. The root cap also produces a secretion to allow RAM to glide more easily through the darkness. The cells of the root cap are constantly being rubbed off as it travels through the soil and RAM must produce more cells regularly, supplying itself with fresh hard-hats when needed. As further protection, RAM has a back-up plan in the form of dormant stem cells, sitting just behind it. If RAM is damaged, these cells can take over the job. As soon as RAM has reached a sufficient depth, it branches out, producing a network of roots, each of them with their own RAMs, each protected by their hard hats.

Almost all plants are vascular. They have systems in place much like own vascular system. Just as our circulatory system carries blood, oxygen and nutrients to all parts of our bodies, plants carry water, oxygen and nutrients to all parts of theirs.

To transport water, the plant must create straws – empty vessels through which water can travel. It does this by laying down cells that die almost as soon as they are created leaving behind a space through which water can enter and ultimately travel around the plant. These straws are known collectively as the xylem. The xylem extends through the stem and branches, into every part of the plant including the leaves. In dicots, the xylem straws are arranged in rings. In monocots, the xylem straws are scattered.

A plant's roots absorb water through osmosis. Osmosis is the movement of water across a permeable membrane from a higher concentration to a lower concentration. When you introduce anything soluble to water, the water molecules stick to the solute. Plants take advantage of this tendency for water to behave in this way by maintaining a high level of solutes within their cells. Since the cells have membranes which allow water to pass through, water from outside the membrane is always attracted through the membrane and into the cells. Once stuck to a solute, a water molecule is no longer pure.

But water is all about equality. When faced with a membrane through which it can pass, water will only be happy when the pure water molecules on either side of the membrane are equal. In this way, the root is able to attract water molecules from the soil into the cells within the plant as the $H_2O$ molecules attempt to even things up. Once inside the root cells, the water is drawn from cell to cell by the same process of osmosis until it reaches the xylem.

Water molecules don't just stick to solutes, they stick to other substances too and, once inside the thin xylem tubes, they stick to the walls of the xylem and begin to climb up it. This tendency for water to stick to other substances is known as adhesion. Water molecules also like to stick to each other. That's why, when you drop water onto a level surface, it doesn't shatter, but forms a single, cohesive puddle. Water molecules always huddle together whenever possible. Because the xylem tubes are so thin, the adhesive qualities of the xylem walls are stronger than the cohesive qualities of the water molecules so water is pushed up the xylem tubes, against the force of gravity, maintaining the pressure in the xylem and helping to hold the stem upright. The combined processes of adhesion, cohesion and surface tension is known as capillary action.

But capillary action can only push water up into the xylem for a relatively short distance. The plant must then pull water up by other means. The leaves must 'suck' on the xylem straws to allow the water to complete its journey and they are able to do this through transpiration pull. Men sweat, women perspire and plants transpire. Just as sweat is expelled from our bodies through pores, so water is expelled from the skin of the plant through pores on the underside of the leaves known collectively as the stomata.

The stomata's main purpose is to allow carbon dioxide and oxygen to enter the leaf and, as part of the photosynthetic process, the way by which spare oxygen leaves the plant and re-enters our atmosphere. The stomata must be open to receive carbon dioxide. Because its pores are open, the plant suffers loss of water through evaporation. When water is lost through the stomata then water in the soil, being so predictable, enters the plant to make up for the loss. Each drop of water lost through the stomata pulls a replacement drop of water into the xylem straws and so the cycle continues. Transpiration helps to cool the plant just as sweating helps to cool our skin, but it also makes sure the supply of water remains fresh.

For our youngster to survive, it must have a regular supply of water so water lost through transpiration can be replaced. Like us, the warmer it gets, the more the plant sweats and the more water it needs. It follows then, that the bigger the root system, the more chance it has of coming into contact with water and, the bigger the plant, the greater the distance has to be taken up by the root system, both to anchor the plant and to reach a good supply of water. To increase water absorption, tiny hairs grow above the tips of the roots. These root hairs increase the surface area immensely and are generally short-lived. Depending on the species of plant, they might live for a few hours, days, or weeks before they have to be regrown.

To further improve the chances of reaching water, the roots of some plants form a partnership with fungi in the soil. Around 95% of all plants are able to do this and this type of relationship may have existed more or less since plants first evolved a root system. The fungus surrounds the roots, forming a hair-like structure capable of extending the plant's ability to uptake water by a significant amount. It may also play a role in protecting the root system. In return, the plant shares with the fungus some of the sugars it manufactures through photosynthesis. It's a bit like the relationship we have with our dogs – the fungus fetches additional water supplies and gives some amount of protection, while the plant shows its gratitude by feeding the fungus.

## Heard it through the fungus-vine

Plants really can talk to each other. If they're not so far apart, they can do this above ground by using airborne volatile organic compounds. These VOCs are coded messages that travel between plants, usually in gaseous form, and allows them to exchange important information such as the presence of something threatening. But what if plants want to make a long-distance call? Even if the parts of one plant are not in direct contact with the parts of another plant, they can still have a conversation and can even agree to help each other out.

One tree standing on its own is relatively vulnerable. When it's part of a group of trees, it has a far better chance of survival. The first land plants realised the importance of staying close to one another and all the plants since have never forgotten this. Woodland plants know there is safety in numbers, so they indulge in co-operative behaviour. To co-operate they must communicate and they do this by means of their fungus network.

The roots of one tree can talk to the roots of another by using their respective fungal partners. If something happens a plant thinks the others need to know about, messages are passed along the line via the fungus-grapevine and the gossip machine gets to work. If an infestation or a disease is allowed to spread through an entire forest, this could be very costly for all the plants living there. So, when a plant finds itself under attack, it wants the other plants to know about it so they can all be prepared.

But this underground telephone network is about more than delivering bad news, mature trees can volunteer to assist younger ones and deliver help through the fungal network. It is thought that older trees might give away nutrients to young trees in order to help them through their teenage years. There may even be a possibility that older trees can recognise their own offspring and favour them over others. Who knows if the young trees request this assistance or if they receive it by accident simply by being connected to the network? I like to think they make a call. "Hey mum! Give me some carbon, will you?" It certainly gives a whole new meaning to making a trunk call!

Trees aren't the only plants which communicate through their fungus network. Though communication between trees has been fairly widely studied, it is fair to assume that other plants capable of forming relationships with fungi, are also able to plug into the fungal network. So we should probably all think about that when we're handling plants – nothing we do goes without being noticed and we may be the subject of more gossip than we know.

## Successful transplantation

Transplanting pot-grown plants is easy. All you have to do is water the plant first, then carefully remove it from the pot. Because the roots are normally curled around the shape of the pot, you'll get almost all of them without damaging the root tips. If planting in the ground, dig a hole the same depth as the pot but a few inches wider all the way around to encourage the roots to branch out. If moving to a bigger pot, the pot should be just a few sizes bigger than the original for best results.

After transplantation, you must water the plant well. This is not so much to give it a drink but to settle the loose soil around the roots so they're in immediate contact with the soil.

Transplanting field-grown plants (plants already planted into the ground) is a little bit harder. In almost all cases, you'll be lucky to take every one of the root tips with you and even luckier if you don't damage some of the delicate root hairs. Take as much soil with you as you can because you'll have a better chance of taking some parts of the plant's fungus buddy with it. For plants with a dominant RAM – i.e. a thick, central tap root - it's important you dig down deep enough to take the dominant RAM along with the rest of the plant. This will cause the plant far less stress because it won't have to cope with the loss of its leader as well as the stress of being moved. For shallow-rooted, soft-stemmed plants and smaller shrubs, you might achieve this fairly easily but, for larger shrubs, you probably won't.

Transplanting a large shrub can destroy up to 75% of its root tips and will almost certainly result in the break-up of the shrub's marriage to its root-fungus, at least temporarily. As you might imagine, this will stress the plant no end and could even kill it. In all cases, you can reduce the stress on the plant by carrying out transplantation before it wakes up completely from its winter rest. That way, it can divert its energies into re-growing the vital root system before it has to get on with the job of producing fresh leaves, stems and flowers. Early spring, just before the plant breaks dormancy is the best time to transplant almost everything, especially shrubs and trees.

For every plant you intend to move, water it thoroughly a couple of hours beforehand to ensure the xylem is full of water.

If you are confident the root ball is small enough for you to be able to dig out most of it, you can do the job all at once. Get a really sharp spade so, if you do cut the roots, it's a nice clean cut. Dig a trench around the shrub to allow you to position your spade right under the roots. Carefully slide your spade under the roots and cut round and round, going deeper each time until you sever all the roots. It's much kinder for the plant to be cut in this way rather than to pull on it, ripping it from the ground.

If you've got a really big shrub or tree, preparation begins the year before. In spring, cut a trench around the plant, severing some of the side roots and then backfill the trench with light compost. With the tap root still intact, the plant begins to repair the roots you have severed by producing new root tips and roots hairs. Since these new roots will be near the surface, you'll be able to take them with you easily when you move the plant. For the next year, make sure your plant gets enough water and doesn't dry out. The following spring, you will have to sever the older roots but, in allowing it to produce new root tips near the top of the root ball, you will at least give it a chance to recover.

In every case of transplantation, the plant diverts most of its energies into root repair so give it a good prune above ground to reduce the load. It should be placed in a hole the same depth as the original but a greater width. Make sure the soil is firmed well around the roots and there are no voids between the roots and the soil. Any voids you leave will fill with water and could drown the plant so it's important you do this.

Finally, water well and cover the surface of the soil around the plant with home-made compost. The compost will serve to lock in moisture and there's also a good chance it will contain new potential fungus buddies.

Don't forget, what you've just done to the plant is entirely against Nature. Plants don't naturally uproot themselves and hop around to a better position, so keep an eye on it for a year or so. In the first instance, it will almost certainly wilt as pressure in the xylem is lost but it should recover in a week or so. Keep it well-watered but don't drown it. Generally, it rains enough in spring to keep the soil moist but, if summer brings a period of very dry weather, you should bear in mind your transplant may not be able to cope. It's not enough to water the first few centimetres of soil at the top, you need to make sure the water reaches right down to the root tips so keep the hose or the watering can on it for a while.

Don't think you're doing your plant a favour by giving it some chemical fertiliser. It's going to be taking a bit of a rest until it has repaired its root system sufficiently to support more top growth, so leave off the fertilisers for the first year.

After a year, if your plant is doing fine, you can assume it's fully recovered, sign it out of intensive care and go back to treating it the same as all the others.

# Surviving Drought

We know water is an important part of the photosynthesis equation. The xylem extends, like a network of veins into all parts of the plant but the xylem must be filled with water or the network collapses. The vascular bundles - the xylem and phloem together - are the bones of a plant. A plant pumped up and full of water is said to be turgid. Any loss of turgidity for a sustained period of time and it falls over so, in order for plants to remain turgid, the roots must absorb water at the same rate as the stomata lose it. Each morning, plants respond to the rising of the sun by beginning the process of photosynthesis anew. During the main growing season, there are no days off for plants and they need to go to work just like the rest of us.

By midday, the roots are often unable to keep up with the amount of water being lost through the stomata due to transpiration and the plant shuts down production for a while to allow a brief period of recovery. Photosynthesis is halted and the stomata close to prevent further water loss then, a while later, it's all systems go again. So, even plants know their rights and make sure they take their lunch breaks.

Plants are governed by the law of limiting factors. Put simply, this is the same as in a factory where production must stop if one or more component parts are missing. The same applies to plants. If there is not enough light, production must halt. If carbon dioxide levels become low, for example when plants are crowded together in a greenhouse, it doesn't matter how much light and water you give them, they will slow down or stop until more carbon dioxide is available. Without water, the factory shuts down.

In periods of drought, plants take longer breaks. All plants are perfectly able to cope with short periods where water is less available but, if the drought is extended and the soil dries up around the roots, a plant begins to worry. Photosynthesis halts altogether and further moisture loss results in the xylem shrinking.

If there are any flowers or fruit, these are quickly discarded – it simply requires too much energy to keep them going. The veins in the leaves collapse next, the leaf loses its structure and wilts. With photosynthesis halted, the leaves lose the green colour associated with absorption of light, eventually turning yellow and, in severe cases, dropping off altogether. At this stage the plant is under major stress. It is literally dying of thirst but is still in a recoverable situation. Even if the water loss is so severe that the stem collapses, the plant may still recover. Eventually, however, if the roots dry up completely, the delicate balance maintained through the process of osmosis is irredeemable and the plant dies.

Because of the unpredictable nature of our weather, most plants are able to cope with periods of drought. In fact, they pretty much expect to encounter a dry period in summer. Watch your lawn turn from lush green to pale yellow as summer progresses and you'll see what I mean. As autumn approaches and the rains become more frequent, it greens up again and all is well.

In places where there is very low rainfall, evolution has stepped in once again to equip plants with some clever coping mechanisms. In temperate climates, many plants lose their leaves in winter, but some plants growing in areas subject to long periods of drought do this at the height of summer instead. The result is the same; the plant gets a rest and is able to survive through a season that might otherwise kill it.

Other plants, for example the succulents, have developed ways to store up water for use when required. They also have a reduced number of stomata to prevent too much water loss. Some succulents store water in their leaves and you can spot these straight away by their thick, fleshy leaves. Jade plants *(Crassula ovata)* do this, as do *Aloe veras* and many more. Succulents often have a distinctly waxy coating to further prevent moisture loss. Other drought tolerant plants are capable of storing water in engorged roots.

Cacti store water in their stems. The leaves are modified into sharp spines or soft down designed to capture moisture droplets from the air and, in the case of the spines, to protect the plant and its precious water store from grazing animals. If you are ever lost in the desert however, don't cut open a cactus and expect to find water you can drink. For a start, the water is stored in the flesh or sap and this is usually acidic and bitter to taste. Many cacti are poisonous so you need to know exactly what you are doing before asking a cactus for a drink.

Very many of the drought tolerant plants have evolved an extra adaptation to help them cope with low water levels and high levels of sunlight. They may, like sugar cane, be C4 plants. In C4 plants, their ability to absorb and store carbon dioxide more efficiently means they don't have the same carbon dioxide demands as their C3 counterparts. Therefore, the stomata can be open for a reduced amount of time and water loss through transpiration is kept to a minimum.

Some C4 plants have taken this adaptation a stage further. Known as CAM plants, they are so good at storing carbon dioxide, they can open their stomata at night and collect carbon dioxide when there is less risk of moisture loss. They store the carbon dioxide within malic acid, which is why cactus flesh has an acidic taste. When they're ready to harness the energy from the sun during the day they can extract the $CO_2$ from the malic acid and carry out photosynthesis without having to open the stomata. In this way, they can protect themselves when the heat of the sun is likely to increase loss of water through transpiration to dangerous levels – very clever indeed.

## Dr. *Aloe Vera*

*Aloe vera* is a drought-tolerant succulent with long, sword-like leaves full of stored moisture in the form of a thick gel. Cut open an *Aloe vera* leaf and you'll see the clear, sticky gel inside. *Aloe vera* is a CAM plant and so able to store more moisture in its leaves without losing any through open stomata during the day.

The gel inside the leaves is packed with amino acids and has been famous for thousands of years as a miracle-cure for all sorts of ailments. Cleopatra is said to have used it to soften her skin and the plant has been depicted on early Egyptian stone carvings. Certainly, there's no doubt that rubbing A*loe vera* gel fresh from the plant onto minor burns and insect bites can soothe the skin and, for many years, households have kept *Aloe vera* plants on hand for this reason, often passing their plants on from generation to generation. I always have one or two plants for just this purpose and regularly give them as gifts.

Recently, however, *Aloe vera's* popularity has risen with all manner of claims as to its miracle curative properties. Cosmetics companies add it to make-up, cleansers, moisturisers and shampoos and there are some who drink the juice of aloe believing it can treat everything from diabetes to cancer. Certainly, if you are in need of a good laxative and fond of a bitter taste then, by all means, drink *Aloe vera* but, as for many of the other claims, they have yet to be properly proven. In fact, aloe has been shown in some cases to actually slow down the healing process in deep wounds.

Having said that, I've a bit of a cheek questioning the authenticity of a plant whose Latin name *'vera'* means true or genuine. In a hot, dry climate, any damage to leaves causes potentially deadly water loss so the plant must act speedily to remedy the situation. Aloe gel forms a seal over the wound preventing moisture loss and appears also to promote fast repair. It would not have taken the ancient Egyptians long to notice this and so began Dr. *Aloe Vera's* enduring reputation as a healer.

Whatever the effect of *Aloe vera* gel on humans, the plant itself manufactures and stores the gel so it can access water and nutrients in times of drought stress. As well as a host of nutrients, the gel also contains plant growth hormones and we can use the gel to provide a boost to our other plants. The plant hormone salicylic acid controls the stress response, helps plants cope with all kinds of attacks and is present in fairly high amounts in *Aloe vera* gel. Cut out a small piece of the gel and dissolve it in water to make a foliar spray or a root drench that will protect your other plants.

## The fantastic Baobab

If you are ever walking through certain parts of Africa and find yourself in need of a drink, have a look around for a tree that looks like someone's ripped it up and planted it upside down.

The Baobab or 'Upside Down Tree' is arguably the largest succulent in the world and is a master of drought tolerance. For a start, it avoids excess moisture loss by dropping its leaves at the height of summer. In fact, it is without leaves for nine months of the year thus giving rise to the belief the tree is upside down because its leafless branches look like roots. The Baobab is also capable of storing large amounts of water – up to 120,000 litres – in its massive, bulbous trunk.

Local people hollow out part of the Baobab's fibrous stem and fill it with water to use the tree as a travellers' drinking fountain. They then drill holes into the hollowed out stem and plug them and, if anyone needs a drink, they can simply pull out the plug. The Baobab doesn't seem to mind being used like this at all and will continue to grow just the same.

In fact, the Baobab is so accommodating, its trunk has been used for many things. With a girth of up to 10 metres and a life-expectancy of around a thousand years or more, Baobabs have been put to all sorts of uses – as a home, a pub, a bus stop, a prison and even as a toilet complete with flushing facility. Having a door cut into it and even being used as a toilet would kill almost any other tree but it doesn't seem to bother the Baobab. It carries on growing regardless.

Hard to kill as they are, when Baobabs finally do die, they don't hang about and rot slowly like most trees; they simply collapse into a pulpy heap.

## Plants with hangovers

If you've ever had a bad hangover, you probably wouldn't want to wish that misery on anyone, or indeed any*thing*. Much of the pain we suffer from the results of drinking too much alcohol is due to the effects of dehydration. When we drink alcohol, it triggers the suppression of a hormone resulting in our bodies sending water directly to the bladder instead of allowing it to be absorbed by other parts. The morning after, we are in dire need of water so our bodies have to steal water from the brain, causing the brain to shrink and pull on the membrane that attaches our brains to our skulls, resulting in a pounding headache.

Other symptoms you might experience because of the dehydrating effects of excessive drinking are dry mouth, sleepiness, thirst and dizziness. So, if you know how badly it feels to be dehydrated, think how it must feel for your plants. Don't give your plants a hangover if you can possibly help it – you wouldn't like it much if it happened to you.

If you are planning on having a few too many, then a CAM plant might just be your saviour – at least in terms of some of the symptoms. It would seem that the prickly pear cactus is more immune to a hangover than most and it can help us out as well. Prickly pear cactus extract is one of only a few hangover cures that has undergone clinical testing. Testers found it may well half the severity of your average hangover and reduce three symptoms – nausea, loss of appetite and dry mouth. Unfortunately, you'd have to know you were going to actually have a hangover beforehand because the 'cure' only really works if you take it before you start drinking.

## Reviving an under-watered pot plant

Plants growing in pots are more likely to suffer from the effects of moisture-loss than those in the garden. It's not as if the roots can go in search of additional water so, if you forget to water, the plant has pretty much had it. If you catch it on time, you can save it but, for a severely under-watered plant, it's not enough to apply water in the usual way to the top of the pot. If the compost is very dry, it will have lost its ability to soak up any water and the water will, in all likelihood, pass right through.

What you need to do is to get water to the root tips and to do this you must fully hydrate the soil. Simply immerse the whole pot in water and leave it there for an hour or two until the soil is fully moist again, then allow the excess water to drain and your plant should recover in a few days.

Don't get a guilt complex about your neglect and smother the plant with affection afterwards. You must wait until the soil is relatively dry again before you water the plant or you risk drowning the plant and killing it with kindness.

# Not Waving but Drowning

Strangely enough, as dependent on water as they are, plants are generally a lot more tolerant of drought than they are of drowning. Too much water is not a good thing at all for most plants.

It is often said that plants breathe in carbon dioxide and breathe out oxygen during the day and then, at night, they do the opposite. This common misconception gives rise to the age old advice that you should never keep houseplants in your bedroom. However, you needn't worry about plants stealing your oxygen as you sleep any more than you do when you're awake.

Humans, other animals and plants all respire for the same purpose – the intake of oxygen helps to break up stored sugars which are released as carbon dioxide, water and energy. The more energy required, the more oxygen must be taken in. Think of how our breathing quickens when we run and you'll get the picture.

During photosynthesis, plants take in carbon dioxide and use it to form sugars. Plants, with the help of oxygen, are able to release the energy from sugars slowly as they go about their day to day business of cellular division, growing, flowering, etc. Plants therefore intake oxygen (respire) both day and night. Respiration in plants is the opposite of photosynthesis – oxygen plus sugars equals carbon dioxide plus water and energy.

Unlike us, plants don't have lungs to breathe. Instead, oxygen passes through the cell walls in all parts of the plant – the stems, leaves and roots, with the roots absorbing oxygen mainly from air pockets in the soil. When soil becomes waterlogged, the air pockets in the soil are filled with water so the amount of oxygen is reduced and the roots begin to suffocate.

The plant's response to this attack on its roots is more or less the same as when threatened by drought. This is down to the Law of Limiting Factors. Even though there is an excess of water, the roots cannot uptake the water because they cannot take in the oxygen they need to perform daily functions. As soon as the roots communicate to the rest of the plant they are struggling to breathe, the stomata closes and photosynthesis halts.

Just as with drought, the leaves turn yellow and wilt as they suspend production of sugars and wait for the soil to dry out, for oxygen in the soil to become available again and for the roots to begin working properly. If this happens within a relatively short space of time, the plant recovers but, if the situation continues it gets even worse.

As a temporary measure, plants can respire without oxygen, a process known as anaerobic respiration. When humans work out too heavily, they use up more energy in the muscle than can be replaced by the normal intake of oxygen. When this happens, the body converts energy inefficiently by producing lactic acid and this creates the burning feeling you get in your muscles when you over-work yourself. Eventually, if you continue to work under oxygen deficiency, your muscles will stop working altogether. Plants, under low oxygen levels can continue to 'do work' by producing ethanol and it is anaerobic respiration in a fungus we have to thank for our beer and wine. The yeast consumes the sugars from the plants but, when subjected to low oxygen levels, uses anaerobic respiration and produces alcohol as a by-product.

Just as with us, anaerobic respiration is so inefficient that plants can't keep it up for long. A plant may be able to continue respiration without sufficient oxygen for a few days but it will pay a high price because the parts of the plant involved in anaerobic respiration eventually burn out altogether. Roots without a good supply of oxygen suffer a great deal. On top of all this, excess moisture in the soil encourages the growth of harmful fungi which attack the weakened roots, causing them to rot. Eventually, without the support of its roots, the plant dies.

Outside, plants are usually able to cope with temporary periods of flooding by shutting down the factory until conditions are good enough to start it up again. Indoors, it's a different story. This is partly because plant owners often think the effects of over-watering have been caused by under-watering, since the symptoms are so similar. The leaves turn yellow, then brown, the plant wilts and the well-meaning plant owner waters the plant, making the situation a whole lot worse.

# How to save an over-watered houseplant

Over-watering of houseplants is a relatively common non-pathogenic disorder. Our houseplants sit in the corner for a while, forgotten and ignored. When they begin to complain about it, we get that guilty feeling and lavish them with kindness, vowing never to forget to water them again. Sadly, this often means we accidentally drown them. Unfortunately, by the time your houseplant has told you its drowning, it's already under severe stress so you'll have to act fast. Even with treatment, the survival rate for over-watered plants is low, but you owe it to the plant to try.

If you can get it out of its container, remove

it from the pot and place it on newspaper to soak up the excess water. After a few hours sitting on the paper, re-pot the plant into a bigger pot where its roots can access oxygen from the new potting medium.

Remember, the cause may be excess water but the problem is actually the lack of it because the roots can't function. Some parts of the root system may already be dead so the plant's ability to photosynthesise will be severely compromised. Move the plant to a shady location to prevent further moisture loss from the leaves and let it take its time to get back to work. Don't let the soil completely dry out or you'll make the problem worse.

Plants recover very slowly from over-watering so you'll need to give it some TLC for quite a while. Mist the leaves regularly, don't let water stand for any length of time in the drip tray and don't feed it until you're sure it's fully recovered.

The best cure for over-watering is prevention so you should always empty drip trays under houseplants after watering. I water most of my houseplants by putting them in the shower and turning the cold water on for a while. Then I turn the shower off and let them sit in the shower tray so the excess water can drain away. The next time I water them is when the soil feels dry. Plants watered in this way rarely suffer from drowning.

# Stop your garden plants from drowning

Plants grow less in winter so they respire less. Like the difference between you and I sitting on the sofa or running for a bus, they need to take in less oxygen to perform their winter tasks. So, when the likelihood of excess rainfall and flooding is greater, plants can cope better with oxygen loss because they need less oxygen in the first place. In the garden, problems occur when the soil retains moisture at a time when plants needs to respire the most so, if you have permanently waterlogged soil, you need to take action in order to grow the most diverse range of plants.

If you must have a lawn in an area prone to water logging, then your only option is to install land drains beneath the lawn. Installed correctly, the land drains remove excess water away from the grass roots and you'll be amazed at the difference. Land drains take advantage of the tendency for water to want to even things out. Faced with an empty, or near-empty, space such as that maintained within land drains, the surrounding water flows into this space to even out the number of water molecules. Water is pulled out of the soil into the drains and once there, it is encouraged to run downhill towards a drainage point.

If you have a naturally high water table in your garden and you absolutely cannot solve the situation, then you should consider growing most of your plants in raised beds. Growing in raised beds lets you control the nature of the soil and allows you to grow your plants above the natural water table.

# Coping with Wet Feet

Even though being permanently immersed in water may be deadly for most plants, some positively thrive on it. Almost all plants living in, or under, water have special adaptations to prevent suffocation of roots. They might have air spaces in the stems which allow them to absorb and store oxygen from the atmosphere and transfer it down to the roots. Many use these air spaces to give them increased buoyancy so they can float around on the surface or float up to the top of the water where there is more available light.

The aquatics are masters at swimming, diving, floating or wading depending on where they find themselves.

On the edges of water, we find the marginal aquatics or emergent plants. These plants have their feet in water and the rest of their bodies out of it and, of all the aquatics, they are most like ordinary land plants. In fact, many are perfectly capable of surviving outside water with just the occasional flooding every now and then and they happily grow in boggy soil. The marginals have tissues in the roots which contain large spaces. When the roots are submerged in water, they can draw oxygen from the air and deposit it in these spaces so the roots have oxygen in order to respire. At times when the pond water retreats and they find themselves on dry land, they can revert to obtaining oxygen for the roots via the usual way.

With their roots in water and their leaves floating on the surface, we find the deep water aquatics such as water lilies. They have most of their bodies in water and use leaves on the surface to capture sunlight. Since water loss is not an issue, their stomata do not require the protection achieved through being placed on the underside of the leaves, so the stomata are located on top of the leaves. Their stems are less rigid than other plants since they are supported by water. If you take deep water aquatics out of water, the stems simply fall over so they're not like marginals and absolutely must have water to support them. However, like marginals, their roots can receive oxygen through specially adapted air spaces.

Submerged aquatics live completely under water. With the quality of sunlight reduced, they have to cope with very low light levels so have all the adaptations you might expect from a shade plant except that, rather than having large solid leaves, their leaves are usually heavily dissected. This is to prevent damage from water currents and increase the ratio of surface area to volume. For any healthy pond or water course, these submerged aquatics are essential. Their photosynthetic activity releases excess oxygen into the water, often seen as bubbles floating up to the surface. Release of oxygen like this regulates the amount of dissolved oxygen available for other inhabitants such as fish.

Right on the surface, we find the floating aquatics. These plants have adaptations such as air-filled leaf stalks to make them highly buoyant and they can float about on the surface of water like little rubber ducks. By staying afloat, they can easily get access to carbon dioxide and sunlight and they don't need roots for anchor. Their much-reduced roots absorb nutrients directly from water. By shading out any plants below them, they can even get rid of some of the competition. They can, however, reproduce rapidly and this presents a real threat for any submerged aquatics underneath a thick blanket of floating aquatics.

A healthy freshwater ecosystem usually has all four types of aquatic plants living within it – marginal, deep-water, free floating and submerged. They are highly adapted to their chosen habitats and are often surprisingly complex. Sophisticated as they are, they can find themselves under threat from the least well-developed plants of all – the algae. Just as they did millions of years ago, algae can rapidly multiply in a water system, sometimes using up all the available oxygen and causing a mini extinction event.

# Amazing snorkelling 'mammalian' mangroves

Mangroves are trees or shrubs which grow in the muddy, oxygen-poor margins of tropical coastal regions. They provide a valuable service to their human neighbours, protecting the coastline from erosion and presenting a physical barrier from storms, but a mangrove swamp is one of the harshest environments a plant could possibly find itself in. Not only do they have to cope with permanently waterlogged roots but, each day when the tide rises, they are inundated with salty water. Just as we cannot drink salt water, neither can plants. In fact, salt is highly poisonous to most plants and mangroves must get rid of the salt or prevent it from entering them and killing them.

Mangroves have ultra-filtration systems, allowing them to filter out most of the salt before it enters the rest of the plant. When salt does get through, it is sometimes diverted to the older leaves which the plant then ditches – this is kind of the same as we might suck out poison and spit it out. A few species of mangrove are able to tolerate high salt levels in their stems and leaves which they then secrete though special cells on their leaves. You can actually see the crystals of salt on the undersides of the leaves.

An abundance of salty water means a

shortage of fresh water and so mangroves have to conserve as much water as they can. Unlike most plants they turn their leaves away from the sun to avoid drying out. Some have succulent leaves which are able to store water. Thus we have a group of plants that has had to evolve adaptations allowing it to survive in drought and in waterlogged conditions at the same time.

Mangrove tree roots have to live in a stinky, oxygen-poor, mosquito infested mud which is very low in nutrients. Some mangroves push themselves out of the water on stilts so the roots can breathe easier. Others develop modified roots (pneumatophores) that grow upwards and out of the water like snorkels, allowing the roots to take oxygen directly from the air. If you are ever lucky enough to encounter certain mangrove trees, you will see them surrounded by their snorkel-like roots, sticking up out of the water.

Life is harsh enough for a mature mangrove tree. A fully-developed mangrove tree can afford to shed some leaves to get rid of salt and it has had time to develop its stilts or its snorkels to help it breathe. But, as you might imagine, it's a lot harder for a youngster to survive. To increase their offspring's chances, many mangroves have adapted a feature more commonly attributed to mammals – they 'give birth' to live young. This is nothing like the plantlet clones produced by plants like strawberries, brambles and spider plants. The young of these mangroves are genetic individuals, conceived of two separate parents.

The mangrove seeds germinate whilst still attached to the parent and grow into leafless seedlings which continue to be fed by mum. Not until they are toddlers are they released into the big wide world. Some mangrove seedlings will drop off their mum and embed themselves in the mud close by. They will spend the rest of their lives close to their mothers, surrounded by family, but there won't be room for every one. Those seedlings released from their mothers when the tide is in, are destined for a journey across the sea. The seedlings are equipped with a flotation device in the form of air stored in a photosynthesising 'stem'. Besides the packed lunch provided by their parents, they have the ability to photosynthesise for themselves, so young mangroves can survive for long periods of time at sea. They set out floating vertically, like they're wading out. In deep water, they lie on their backs. In shallower water, the youngsters' buoyancy is once again reduced and they switch back, from horizontal to vertical. This increases their chances of catching on mud and helps them hold onto the shallow sea bed like an anchor. Dehydration triggers the production of roots and the plants are able to establish themselves, in some cases, many miles away from their mums.

# Create a garden pond

Garden ponds are major draws for all sorts of friendly wildlife. Toads and frogs will visit your pond and vacuum up some slugs along the way so it's well worth having a pond for that reason alone. Your pond also acts like a mirror reflecting the plants and features around it and, of course, it's a good excuse for you to grow some aquatic plants.

Locate your pond away from trees if you can or you'll be constantly pulling leaves out of it in autumn. The best place to site a pond is in a well-lit location but away from direct sunlight and, if you can, where you can see it from a window so you can watch all the wildlife approaching it. It also helps to put your pond on a level site.

Once you have decided on the shape and size of the pond, get digging but, in order to avoid the hole filling up with water before you've had a chance to put in the pond liner, give yourself a whole day to do the job, from start to finish. As you are digging, you need to create a slope or steps on one side so, if any animals fall into the pond, they can scramble out again. The last thing you want is to wake up in the morning and find a dead animal floating on your pond. You should also create shelves near the surface for growing marginal aquatics. Make sure there is sufficient depth in the centre of the pond because deep water aquatics such as water lilies need a depth of at least 45cm.

Check the water will sit level in the pond by placing a piece of wood across the hole you have dug, from one bank to another. A spirit level sitting on top of the wood will tell you if the banks are level. Make all the adjustments you have to before you fill with water.

When you are happy with the way your pond looks, remove any sharp stones from the soil in the base of the pond and from the walls, as these might pierce the pond liner when the weight of the water pushes down on it. After you've done this, line the whole thing with a good layer – 5cm or so- of sand to provide extra protection for the liner. You can use special pond underlay instead of sand but, as an extra precaution, I would actually recommend using both sand and underlay because a pierced pond liner is the last thing you want. Next, you lay the liner over the top and allow it to settle over the hole, making sure you smooth it out as much as you can, reducing the number of folds.

To buy the correct size of pond liner, use this calculation:

*Length= length of the pond at its longest point, plus 2 x the depth, plus around 60cm to allow for 30cm either side to lie over the rim of the pond.*

*Width = width of the pond at its widest point, plus 2 x the depth, plus around 60cm.*

Slowly fill your pond and allow plenty of time for this. It takes longer than you think to fill a pond, even with a hose. As the pond fills up with water, fold the liner neatly and, once filled, leave the pond overnight to settle before edging it. How you edge the pond is up to you – you can use rocks, paving slabs or turf, but try to make it look as natural as possible.

Leave your pond for around a week before you plant it so any chlorine in the tap water will come out. Plant it up with marginals on the shelves, deep water aquatics dropped into the pond and a small amount of submerged plants for oxygenation, plus floating plants to provide shade and cover for pond dwellers. Deep water aquatics and marginals should be potted into special mesh pots to allow the roots to spread through the holes in the pot and you need to use compost specially designed for aquatic plants. Ordinary compost (especially one containing peat) is too light and simply floats up to the top, plus it often contains too many nutrients, encouraging rapid algal growth. It is possible to use plain gravel for planting deep water aquatics too because the plants get most of their nutrients from the water and not the soil.

Algae multiply quickly when presented with plenty of sunlight, so you can reduce the presence of algae in your pond by shading out the surface of the water with pond plants. As a general rule, somewhere around 50 per cent of the surface of your pond should be taken up by plants.

Don't forget your pond is a mirror so plant something great beside it so it can be reflected in the water. A nice Japanese *Acer* looks fantastic with its coloured leaves mirrored in the pond or a *Rhododendron* with its bright flowers.

# Part Six: We all have to Die Sometime

# Annual, biennial, ephemeral

Even though plants keep undifferentiated cells through most of their lives and even though some can easily regenerate when they are damaged, there comes a time when, just like every other living thing, they must die. Death and old age is as inevitable for plants as it is for us. If those that live in the present do not die, there cannot be room for new life in the future. From the mightiest tree to the tiniest of ground-hugging plants, each has a reasonably predictable senescence, a time when they must grow old and weak, and then die. The age at which natural death occurs depends on the species of plant, on the temperature fluctuations in the country in which they live and on whether or not they have secured the future of their species, bringing into existence the new life they will be making way for when they die.

Though plants can, and sometimes do, die of old age, mortality rates amongst plants are so high, it is rare for a plant to make it all the way to a natural death from old age. If a tree can make it to a good size, it's got a real chance of living to its full life expectancy. A soft-stemmed flowering plant in a home garden may live for many years under the care of its human protectors but the same plant in the wild would not fare so well.

So what is the average life expectancy for a plant? The short answer to that is, there isn't one. Certain plants have a life expectancy of just a few months while others live for years and even centuries. For convenience sake, we group plants according to their approximate life expectancy - that is the time it normally takes for them to germinate, put on vegetative growth, flower, set seed, and then die. To determine which of our plants fit into the various groups, we use the following terms: ephemerals, for those that live for a very short time; annuals for those that live for one season; biennials, for those that live for two seasons and perennials, for those that live for more than two growing seasons.

However, the terms ephemeral, annual, biennial and perennial are just way too simplistic for such a complex and diverse group of living things such as plants. For example, some plants we treat as annuals in cooler climates are actually perennials that cannot cope with low temperatures, and some perennials only live for a handful of years, while others will live for centuries. A better way to think of plant life expectancy is to think in terms of the investment they put into reproduction. To understand plant life expectancy, it is more appropriate to think of plants in terms of two groups - monocarpic and polycarpic.

The usual definition of a monocarpic plant is one that flowers only once in its lifetime but the true definition is of one that produces a crop of seeds only once. The act of flowering alone does not result in reproduction. Only after a monocarpic plant reproduces and successfully ensures the next generation, does it rapidly age and die.

We are arguably at the peak stage in our lives during our young adulthood. By then, we have matured enough and learned enough of life's lessons to enable us to employ a number of survival tactics independently of our parents. We are also at the peak of our reproductive usefulness and therefore, in general, at the very peak of our fitness capabilities. All things being equal and, if we don't abuse ourselves, our cells are healthy and fit, able to regenerate quickly and recover from attack. As we begin to outlive our reproductive usefulness, we start to feel the effects of ageing more keenly. So it is with plants only, while we have many years and many opportunities to reproduce and can, of course, reproduce more than once, monocarpic plants have just one chance. When that one chance is exploited, ageing and death follows quickly.

Whether or not a plant chooses to invest in just one throw of the dice is all down to the behaviour of its SAMs. It would appear that, when a shoot apical meristem invests in becoming a reproductive organ - a flower and more importantly a seed - then that investment is absolute. In monocarpic plants all, or almost all, active SAMs shift from stem elongation to flower production either at the same time or at least in the same season. This means there are not enough SAMs left over capable of regenerating stems.

It seems logical the reason plants should commit suicide in this way is because the likelihood of its survival into a second season is extremely low, for whatever reason. With the odds stacked against making it beyond season one, the plant may as well take an all or nothing approach to reproduction. This is why most ephemerals, annuals and biennials are likely to be monocarpic.

Ephemerals exist above ground for a very short period - a few months normally but sometimes just a few weeks. They tend to be plants whose chances of survival are extremely low or that live in such places where the conditions are only right for growth during a very short time. If you have heard of, or even witnessed, the sudden coming to life of arid landscapes after a rare fall of rain, the plants responsible for this incredible show are generally ephemerals. These desert ephemerals have waited as seeds, some of them for many years, until rain breaks the seed's dormancy and there is a sudden rush to produce more seeds before the land becomes once again too dry for life above ground. And so it is that, after rain, deserts can become a carpet of green, then a mass of flowers and then nothing but dry land again, all in the space of a few weeks.

Some of our common weeds are ephemerals - chickweed, hairy bittercress and groundsel to name a few. For plants that have to compete with other wild plants and to dodge a gardener's hoe or herbicides, it makes perfect sense for them to get in all their living in such a short space of time. The parent plant may live a short life but the seeds can remain viable for many years thus ensuring the species is protected from extinction, no matter how hard we try to remove it.

Sometimes you will hear the term 'spring ephemerals' when referring to plants like *Trilliums* and bluebells but don't be fooled by the name. These plants are not really ephemerals but rather woodland perennial plants that produce vegetative growth and flowers for the short time in between the soil warming up and the leaves appearing the trees. Since they don't actually die after flowering but merely sleep underground for a long time, they cannot be classed as true ephemerals.

It's no surprise many of our seasonal bedding plants are monocarpic. We want as many blooms as possible over a relatively short space of time so we'll appreciate a plant prepared to invest so heavily in flower production, even if it means their lives will be no more than a season long. The annuals and short-lived perennials we use for seasonal bedding are the male spring lambs of the plant world - best enjoyed while they are young and not much use to us beyond a certain age.

It is often said annuals live for one year and this is a good way to describe them, but it isn't strictly true. Since most annuals are killed off by frost, the length of time they live is a quite a bit less than a year, depending on where they live. Annual plants live for just one growing season. During that time, they throw all their eggs into one basket and produce what are arguably the most spectacular displays of flowers in the entire plant world. This is why we make the seasonal pilgrimage to the plant nursery to buy them in their millions, even though we know they won't last very long.

Biennials take a year to prepare themselves for the most important event in their lives. In their first year, they produce leaves and stems, gathering up energy and putting on weight. Like athletes training for the Olympics, they prepare themselves for the flowering event by ensuring they are as strong as possible. In year two, it's an all-or-nothing race to produce flowers and seeds. Again, they choose to invest absolutely in creating seeds and are unable to regenerate afterwards. After two years, they must retire and allow younger, fitter athletes to take their place.

Some monocarpic plants don't truly die after setting seed because they produce clones. The banana plant does this and so do many plants in the bromeliad group. In this way, their clones ('pups') carry on the good work and the DNA of the parent plant is retained.

## Never mind annuals and biennials, how about a centennial?

Century plants are plants in the agave family whose common name derives from the belief they flowered only once every hundred years. Actually they do flower only once but usually at between ten and twenty years old. Still, it's a long time to wait to produce a flower. I am particularly fond of an *Agave* living in Glasgow's Botanic Gardens. Located at the end of each of its massive succulent leaves is a spike that could stab you right through to the bone. I guess if you're going to take decades to flower, you need to protect yourself as well as you can.

Being a lover of plants capable of looking after themselves, I have to admit to smuggling a little variegated *Agave* pup home in my luggage many years ago on my first trip to Spain. This pup lives on my windowsill and hasn't grown all that much in thirty years and I often think the biggest favour I can do for it is to return it to its homeland. One thing is for sure, for as long as I keep it in a restrictive pot, I won't see my version of a century plant flower.

When century plants do flower, they send up enormous flower stalks – some as tall as seven metres. When the century plant at Kew Gardens flowered, they had to remove a glass pane from the roof to accommodate its statuesque flower. After flowering, being monocarpic, these magnificent plants 'die'. Of course they don't really die since they live on as clones in the pups they leave behind.

## Kiss me Hardy

We class annuals as either hardy or half-hardy. We often say hardy annuals are resistant to frost and half-hardys are not. In fact both hardy annuals and half-hardy annuals are killed off by frost after flowering but it is their resistance to frost before flowering that distinguishes them.

Many hardy annuals can be sown in autumn while the soil is still warm enough to encourage germination. The little seedlings grow just a small amount and then sit all through winter, standing up to frost but not prepared to venture very far above the soil. Their roots however have the opportunity to put on better growth than any sown in spring so that, when spring arrives, they can get ahead of the game by several weeks. Cornflowers (*Centaurea cyanus*), Love-in-a-mist (*Nigella damascena*), Scotch Marigolds (*Calendula officinalis*), corn poppies *(Papaver rhoeas)* and poached egg plants *(Limnanthes douglasii)* all respond really well to autumn sowing. Most hardy annuals can be sown straight into the ground where they are to flower so you can scatter their seeds amongst your perennials for an added boost of colour while you wait for perennials to flower.

Half-hardy annuals need to be started off in the greenhouse, usually in spring, and then protected from frost while they are young. Only when all danger of frost has passed should they be planted out and, of course, they must be properly 'hardened off' to get them used to life outside the greenhouse before you finally let them loose. Popular half-hardies are: *Lobelia erinus,* French marigolds *(Tagetes patula), Petunias* and Busy Lizzies *(Impatiens).*

## How to grow great carrots

The carrot is a biennial plant, destined to live for two years, putting on vegetative growth in its first year and flowers and seed in its second. Because we harvest carrots in their first year, the carrot never gets to live as long as it should and doesn't achieve the satisfaction of knowing it has produced progeny. Sometimes though, carrots get the better of us and sneak in a flower in their first year. Gardeners refer to this as 'bolting' or 'running to seed'. To grow good carrots and prevent them from bolting, we must understand what the carrot plant is trying to do.

The domestic carrot's wild ancestors were nothing like the carrot of today. Their roots were bitter and stringy and they were probably white to pale purple in colour. Over time, certain plants were selected for thickness of roots, sweetness and then eventually for the orange colour most now display. For the unfortunate carrot, being bitter and stringy was an advantage because nothing much wanted to eat it. When your root is sweet and full of lovely sugars, it's a distinct disadvantage since you must be dug up in your teens to be eaten.

Like all biennials, in its first year of life, the carrot spends its time collecting energy from the sun, converting it to sugars. It stores these sugars in a single taproot. When it senses a seasonal change telling it a new year has begun, it puts out side roots and begins to use up the stored sugars in the taproot to produce a flower and then to set seed. If you sow your carrots too early and then there is a cold spell, the carrot assumes this cold spell is winter. When the weather warms up again, it thinks a new year has begun and sends up a flower stalk. Once a carrot produces a flower bud it begins to pull the sugars out of the root to help it produce a flower and then seeds. If it is allowed to do this, the root is no longer quite so tasty, so we have to stop it doing so in its first year.

Carrots should be sown once the weather has settled down and the soil has warmed up. If a sudden cold spell occurs, you need to cover the carrots with horticultural fleece to stop them thinking winter is approaching. You can apply this treatment to prevent other biennial vegetables from bolting too – chard, parsley, turnips, spinach and beetroot.

Carrots should not be grown on freshly manured soil because the high nitrogen content in the soil makes the plant lazy about storing up sugars. The tap root stops reaching deeper into the soil in search of nutrients and runs off the straight and narrow so you get forked carrots. It may also begin to produce side roots or hairs earlier than it should. Carrots also fork when they hit a stone or other obstruction in the soil so you should grow them in very light soil which has been sieved if possible. If you have heavy soil, make up a container with sandy compost and grow your carrots in that. To stop your carrots from going green at the top, mound some soil around the bottom of the leaves so the root won't see any sunlight.

Carrots make their sugars best during moderate weather. If you grow them through a long, hot summer, their roots go woody in preparation for adulthood so it's best to grow and harvest them either side of the very hottest month in your area. Finally, once your carrots are growing well, keep to a regular watering regime. If the soil gets too dry, the root starts to produce lots of side roots in an effort to find more water. If the soil gets too wet, the root sometimes splits right along its length.

# Soft-stemmed perennials

Polycarpic plants are long-term investors. Rather than invest their reproductive fortunes in a single spin of the wheel, they take a more cautious approach, building up their stocks of progeny over a period of years. Each year, they set aside some of their SAMs for the production of flowers and bank the rest. This SAM insurance policy allows them to survive for years.

The perennials live for more than two growing seasons. Some can live for just three or four years before they've exhausted their supplies and others, like the giant redwoods, live for thousands of years. There are those that retain soft stems and those that produce thickened, woody stems. Since we like to categorise things so much, we divide them into two groups – soft-stemmed perennials and woody-stemmed perennials.

As winter approaches, a young soft-stemmed perennial faces a new set of challenges. If it has managed to survive being trampled, nibbled on, disease and pest attack it will be in need of a good, long sleep. It must, however, avoid the deadly effects of frost or its first sleep will be its last. When we get frostbite, our skin and the underlying tissues freeze. In severe cases, this damage is irreversible and the tissues affected are lost forever. The same is so for soft-stemmed plants. Freezing affects the circulatory system such that water and sugars cannot flow through the xylem and phloem respectively. The cells of the plant turn to ice and death is imminent. Soft-stemmed perennials must mitigate the effects of freezing temperatures well enough so they can survive winter.

Herbaceous perennials are defined as soft-stemmed perennials that die back in winter. For the most part, this is true even though some herbaceous perennials retain a presence above ground. Of course, the plant doesn't really die back. It simply retreats to its underground home and refuses to go to work until conditions are better. If the above-ground factory cannot function because the conveyor-belt is all frozen up, then the factory must shut down and be completely dismantled, or at least reduced in size with its doors firmly shut.

Herbaceous perennials are genetically programmed to anticipate the coming of winter and to take steps to avoid its deathly grip. All summer long they've been storing up food in the roots – enough to survive winter's siege. As winter approaches, they simply retreat underground and remain there, all snug and warm, until it's time for them to come out again.

When the factory collapses, the shoot apical meristems which controlled all the growth above ground are lost but an herbaceous perennial has banked enough meristematic cells to initiate a full re-growth in spring. It forms adventitious buds near the roots, either at ground level or just below it. These buds are the plant's future SAMs and they will wait until the time is right for them to be factory supervisors in the next growing season.

In most grassy perennials, the stems are permanently located either at, or just below, ground level. What we see above ground is a collection of leaves originating from a very short stem. The leaves of some grassy perennials fall back altogether while others retain some leaves to help protect the SAM. This is why our lawns don't grow in winter but remain green all the same.

Bulbs are soft-stemmed perennials with a particularly large food storage site. They concentrate their food stores not in roots, but in a swollen stem. In the centre of the bulb, is a mini plant, protected by the food store around it. The bulb can then lay dormant underground until the time is right for it to produce both roots and a shoot. This is very convenient for us since we can dig them up at the right time and carry them to other places without fear of damaging the dormant plant inside.

Any plant that has to survive through every season must have a strict schedule. It must send up shoots, flower, collect food and initiate die-back, all according to a job-sheet that rarely deviates from year to year. The best thing about herbaceous perennials is that each species has a slightly different schedule and the predictability of their flowering times means we can plan our gardens accordingly.

If we choose carefully, we can make sure there is something of flowering interest just about all the way through the year, from early spring until late autumn. Herbaceous perennials don't flower for as long as the annuals but, with careful planning, the joy of a large flowering stand of Siberian irises, followed by the disappointment of watching the flowers wither, is quickly replaced by something else coming into bloom a short distance away. Because they don't tend to flower for very long, I think this makes us appreciate herbaceous perennials all the more.

Herbaceous perennials are so good at hiding underground, one of the biggest dangers to them during winter is gardeners. First-time gardeners often think the plant they have bought has died and throw it away. If they don't throw it away, they're going to get a nice surprise in the new season. I always love it when a new gardener is so delighted their plant has come back from the dead even bigger and better than before. Even the most experienced gardener can forget where some of their perennials are located and accidentally dig them up during a winter clear-out. Luckily, if you do dig one up by mistake it's no harm done if you quickly re-plant it.

## Gertrude Jekyll, the artist who painted with flowers

Gertrude Jekyll was a hugely influential Edwardian garden designer who loved her herbaceous borders. A pioneer of the English Flower Garden design style, she disliked the orderly bedding schemes of the Victorian era. She was not one for forcing plants to grow where their owners wanted them to grow but, instead, carried out meticulous research, choosing only plants most suitable to the location.

She was a gifted artist whose failing eyesight caused her to consider planting flowering herbaceous perennials in large drifts of plants so they could be seen as blocks of colour from a distance. Influenced herself by Impressionists paintings, she created big brush-strokes of colour across the landscape and designed her planting schemes by paying careful attention to the colour wheel. She created 'hot' borders with fiery red, yellow and orange flowering plants and 'cool' borders of pinks, blues and whites. Planting just one species of herbaceous perennial plant on its own was rarely an option for her and her philosophy continues to be practised today.

If you want to plant herbaceous perennials, try to plant them in groups. As a general rule, and if you have the space, no less than three together, and always in odd numbers. In groups of three, five or seven is the best way to get the most out of your herbaceous perennials.

## Why is grass so tasty?

Grass, collectively speaking, is just about everywhere. It's in our lawns, in fields, on roadside verges, clothing the sides of hills, from the Arctic tundra to the Savannah plains. Everywhere grass is, there's something that eats it, whether it's our lawnmowers or various herbivorous animals. A whole group of animals – the ruminants – have evolved to cope with the difficulty of fully digesting the cellulose in grass. Of course, the term 'grass' is commonly used to describe a group of grassy plants growing together. Like the animals grazing on them, grass plants themselves seem to prefer to hang about in herds. Different species of grass plants will dominate in different parts of the world depending on their adaptations to climatic conditions but you rarely find one of these grass species growing on its own.

The question of why grass should be quite so tasty kept me awake one night. I wondered why something would accept being eaten to such an extent without employing a more effective defence system. Well, grass does have a defence in the form of abundant cellulose but, when ruminants evolved a different type of digestive system to cope with cellulose, why didn't grasses evolve some other kind of response?

The answer could lie in the place grass occupies under secondary succession. Grass, whether it is in our lawns or in a grazed field, is, on the whole, kept in a state of youthfulness. It is perfectly evolved to cope with being eaten or mown and you could almost say it totally expects to be so. With its SAMs located close to ground level, it can sacrifice leaves to a hungry mouth without risking death but this also means it doesn't have much opportunity to reach sexual maturity and flower.

If you left a field of grass on its own without grazing or cutting it, grass flowers would spring up. Left alone for a number of years this field would soon become the target of broad-leaved herbaceous perennials and they would use the protection of the tall grass flower stems in order to get established easily. Think of how quickly an un-mown lawn degenerates into a mass of buttercups, daisies, dandelions and broad-leaved dock and you will get the picture.

Eventually, the taller herbaceous perennials would out-compete the grasses but the herbaceous perennials would themselves find competition in the shrubs. When the shrubs were tall enough, they would be able to protect the development of tree saplings and, before long, you have woodland and almost nothing is left of the huge herd of grass that used to dominate the site.

So, if grass is allowed to flower indiscriminately without being grazed, it will soon be edged out by bigger plants, not as ideally suited to surviving grazing. That's why we mow our lawns regularly and, though the grass may shout out its annoyance, perhaps it should be grateful we do and maybe that's why it doesn't retaliate.

Even if it's not the real reason why grass remains so tasty to a wide range of animals, for me, it's as good an explanation as any so now I can sleep at night. What would keep me awake is if all the grass species in the world started to play around with their chemistry sets and managed to create a poison to kill all the grazing animals. Come to think of it, that would be a great plot for a movie....

# Propagating herbaceous perennials by division

As we know, herbaceous perennials retain

adventitious buds in order to re-grow their vegetative parts after a winter's rest. This allows them to maintain a predictable height whilst growing in mass by expanding sideways a little more every year. For the most part, if we want them to fill a border, we can leave them alone to do just that. However, in many cases, the longest living part of the plant towards the centre begins to exhibit signs of old-age, producing fewer flowers than the more youthful outer parts. At this time, we can intervene and give the whole plant a facelift to return it to the way it looked in its younger days.

Using a garden fork, lift up the plant and clean off the roots so you can see the most obvious place to make a cut. Often, if you gently pull the plant apart it will break at the most suitable sections, creating two or more new plants, each with roots attached. If you can't pull it apart with your hands, use two forks back to back to force it apart. If this doesn't work and the root system is too entangled you might have to use a knife. I've even had to use a saw at times!

Either way, what you are looking for is a number of good sections, each with buds and a reasonable root system attached. Replant only the sections you have taken from the outer edges and the plant will once again regain its youthful vigour.

If you have any healthy sections left, you can even give them away to your friends or replant elsewhere in the garden. Propagation by division is the easiest way to make new plants and, by the following spring, the new plants will be coming away nice and strong. However, you need to remember that you have just ripped this plant apart and it will need a little bit of TLC. Give the newly planted sections a good water and even cut back some of the green parts to give the roots a little less work to do.

Division of summer flowering herbaceous perennials is best done in autumn after they have flowered and always remove any flower stems before you divide the plant. You should aim to give the new plants around a month of relatively mild weather to settle in and build up some food reserves before they get a good long rest over winter. You can also divide summer flowering perennials in spring but I prefer autumn so they have a recovery period. Spring flowering plants take their rest period in summer and use the summer months to build up new roots so they are best divided in early summer.

Even if you don't need to create new plants, division of most perennials will keep your plant looking healthy and young and should be done every three to five years depending on the plant. Generally, the plant itself will tell you when the time is right.

# Trees and Shrubs

Herbaceous perennials lose all or most of their above-ground parts so they can survive winter. Every year they must build the factories again pretty much from scratch. Because they are limited to putting on above-ground growth for just one growing season each year, they will only ever be able to achieve a certain height. This is helpful to most gardeners and garden designers because their maximum height is so predictable. However, having to put on entirely new growth each year just wouldn't do for trees and shrubs. To occupy the spaces in the canopy that have always been their preserve, they need to grow tall, thick and strong, so they must keep at least some structure above ground upon which to build each year.

Trees and shrubs retain a skeleton of stems throughout the dormant season. In winter, when food is in short supply, soft and succulent stems would be extremely vulnerable so trees and shrubs must create something hard and unpalatable, like wood.

In the stems of dicotyledonous flowering plants and in conifers, meristematic cells are present, capable of initiating lateral growth. Some herbaceous dicots plants can and often do thicken their stems using lateral meristems but all that work is lost when the plant tumbles to the ground in winter. Woody dicots and conifers get to retain their thickened stems, creating wider and stronger stems each year.

In most woody perennials, the vascular bundles (xylem and phloem together) are arranged in neat rings inside the stems – the xylem to the inside and the phloem to the outside. A layer of cells - the vascular cambium - sits in between the xylem and phloem and encourages the growth of secondary xylem and secondary phloem tubes. Each time it does this, the old phloem tubes are pushed to the outside and the old xylem tubes are pushed to the inside. The seasonal lateral growth in most trees and shrubs forms the distinct rings with which we are all familiar.

In the process of creating secondary xylem and phloem tubes, more secondary xylem is created than secondary phloem. The original phloem cells get pushed to the outside and eventually thicken up to form bark or the outer skin of a branch. The old xylem is pushed to the centre and eventually forms the woody structure that will hold the tree or shrub stems nice and steady throughout the life of the plant. In this way, trees can reach great heights with their strengthened woody stems.

If you cut through one of the younger branches of a shrub, you will be able to see the green layer of active xylem and phloem arranged in a circle. To the outside, you will see the old phloem and, to the inside, you will see a much bigger layer of old xylem, responsible for wood. Without these dead xylem cells, trees and shrubs would not be able to maintain a strong enough skeleton and, of course, our lives would not be quite the same without wood.

Winter is harsh on soft green leaves. The dehydrating effects of wind suck out moisture and very strong winds shred leaves to pieces. The freezing temperatures can destroy the cells resulting in frostbite. A tree in full leaf acts like a sail in the face of strong wind and is pushed over far easier. For many trees and shrubs, the choice is clear, to survive winter, they must ditch their leaves.

Commonly, the definition of a deciduous woody perennial is one that loses its leaves in winter but, in areas where the summer months are extremely harsh, they lose their leaves at the height of summer. As well as that, using the term 'loses' make it seem like it's accidental. In fact, deciduous woody perennials deliberately throw away their leaves in a carefully planned process and they do this purely to protect the skeleton throughout the dormant season. As the daily hours of sunlight decrease, our deciduous woody perennial knows winter approaches. In autumn, a layer of cells at the base of each leaf begins to thicken. This is the abscission layer and its purpose is to cut off access to the conveyor belt. The leaves must fall but, at the same time, the wound has to be sealed so water and sugars are not lost from the rest of the plant and no diseases can enter through the wound. Cut off from its water supply, each leaf can no longer photosynthesise and it begins to lose chlorophyll.

As the green chlorophyll is lost, other pigments hiding behind the chlorophyll begin to show – yellow for the xanthophylls, orange for carotenoids, red for anthocyanins and brown for tannins. For a brief period in autumn, these pigments are revealed in all their glory and we get a spectacular colour display. The best autumn colour is seen when autumn's weather is dominated by sunny days and low night time temperatures. Wind or heavy rain causes the leaves to fall before you get to see the colours and a sudden, sharp daytime frost forces the plant to panic and ditch the leaves in double quick time.

Without leaves, deciduous woody perennials go into full factory shut down. Just like herbaceous perennials with their adventitious buds, trees and shrubs have made plans for the following season by producing buds containing all the cells they will need to create new branches and leaves when the time is right. They deserve a well-earned rest and off they will go to sleep, safe in the knowledge that, when they wake up, they will have a structure upon which to build. While they sleep, just like us, they will continue to respire, calling upon the sugar reserves they have stored up to enable them to do so.

While deciduous trees and shrubs fall into a nice deep sleep during winter, evergreens prefer to slow down, switching from full-time work in summer, to part-time work in winter. Conifers are our most well-known evergreens. In winter, they hold onto enough leaves to keep the photosynthesis process moving, albeit a bit more slowly. To protect themselves from the drying effects of the wind and the damaging effects of frost, the leaves of conifers are modified into needles with a strong, waxy coating. Broad-leaved evergreens like hollies and *Rhododendrons* have this waxy coating too.

It isn't strictly true to say that evergreens keep their leaves all year. Instead, they drop their older, less productive leaves all through the year and these leaves, once they hit the ground, take a lot longer to break down due to their waxy protection.

# The cruel practice of tree girdling

Nothing invokes an emotional response quite like the killing of a tree. Because trees are so visible for such a long period of time, we often feel sad when we see them felled. Protesters chain themselves to trees regularly in an attempt to prevent them from being destroyed to make way for our roads and new developments. Sadly, there are times when a tree must be killed for practical reasons. In forests, it may be necessary to cut down one tree for others to have more room to grow. Of course, sick and dying trees must be felled before they fall naturally so they don't cause a danger.

Indiscriminate planting of trees without forward planning is the cause of many an unnecessary tree murder. When we plant a tree, we must consider the impact that tree will have in later life. Research the eventual height and spread of any tree you plan to plant very carefully. If you don't then, many years later, you will force others to make the heart-breaking decision of having to fell the tree because it is too close to another tree, shading out too many other plants or, more commonly, too close to a building. It is selfish for us to enjoy a tree in its young life if it means we will be condemning it to death when it's older.

In my opinion, if a healthy tree must die, the least cruel way to do this is to fell it quickly. A good tree surgeon will take down even the biggest of trees in a matter of hours so I have never understood why anyone would want to cause the long, lingering death of a tree.

Tree girdling involves taking out a circle of bark all the way around the trunk of a tree and then cutting a little way into the trunk, again all the way around, to remove the phloem tissues. This prevents the flow of sugars and literally starves the tree to death. Depending on how it's done, the tree will still be able to make use of its xylem tubes and so it will continue to transport water until, eventually, it runs out of food. It can take more than one growing season for the tree to run out of its reserves and, for all this time, the tree is under considerable stress. A girdled tree can stand for quite a while after it is officially dead and then fall when least expected, so it represents a real danger to the public.

Accidental girdling can occur when a tree is tied to a stake when first planted and the tie is not removed. As the trunk of the tree puts on lateral growth, the tie bites into the bark and strangles the tree, preventing the free movement of sugars and eventually causing the slow death of the tree. Always remember to remove or loosen tree ties as the tree grows.

Girdled trees can be saved by a complex technique known as bridge grafting. Healthy twigs are cut from the tree and inserted into each side of the wounded bark creating a bridge across the damaged area. The tree can re-route its sugars via the phloem in the twigs until it is able to repair the damage itself. This is similar to the way a surgeon might perform bypass surgery on humans.

## How to plant a tree

Right at the entrance to Benmore Botanic Gardens near Dunoon, Scotland, there's a magnificent avenue of Sierra redwoods (*Sequoiadendron giganteum*). It was planted around 1863 by the then owner of the estate – an American, James Piers Patrick. Piers Patrick would have known he would never live long enough to see his avenue in all its current glory but he set about planting it all the same. After the initial planting he would almost certainly have stood looking at his little avenue of saplings and pictured a time, long after his death, when they would achieve the look he wanted.

This is both the satisfaction and the frustration of planting trees. You are planting something that will, in all likelihood, live long after you are gone and you will probably never get to see it the way you imagined it. For this reason, as I said before, you need to be sure your tree will be happy in the place you plant it for all of *its life* and not just for yours. For the same reason, it's not enough to simply dig a hole and shove your young tree in it. Trees must be planted well and with care if they are to be assured a long and happy life.

Trees are supplied container grown, ball and bur lapped or bare rooted. Both ball and bur lapped and bare rooted trees are grown in the field and then dug up to be shipped off to suppliers. In the case of ball and bur lapped, the root ball is surrounded by canvas or by a wire cage. You must cut away the wire and the canvas before you plant the tree or the roots will be restricted.

Water the tree well around an hour before planting. Dig a hole as deep as the depth to which the soil has been sitting on the tree but around three times as wide.

The more a tree gets pushed around by the wind, the more it will work to establish a good anchoring root system and to thicken up its stem. However, you do have to stake most trees for a while, to support the roots while they obtain anchorage. For very young trees, you shouldn't have to stake at all. If you do have to stake, your stake should be around one third of the height of the tree. You can use a straight stake or drive it in at an angle – it's up to you. Generally, a straight stake is used for level ground and an angled stake for slopes. Make sure you position the stake such that the tree won't be pushed towards it in the wind. I drive my tree stakes into the ground before placing the tree in the hole so I know I won't be causing any damage to the roots with the tree stake.

Place the tree in the hole beside the stake. With bare rooted trees in particular, avoid turning the roots in on themselves just to make them fit the hole. If you can't spread the roots out fully, dig a wider hole.

In all cases, what you are trying to avoid is girdling of the roots. This is when the roots, instead of growing sideways, twist around the root ball instead. This can happen when the hole you have dug isn't wide enough or when you restrict the roots by failing to remove netting or wires. Just as when trees have girdled trunks, when the roots of trees are girdled, the tree's life is drastically shortened because it strangles itself. On top of that, it won't have side roots to stabilise it and will ultimately fall when the canopy gets too heavy for it.

Once you are sure the roots have plenty of room, back-fill with the soil that was originally in the hole. The general consensus these days is not to put fertiliser in the planting hole. If the soil closest to the tree roots is more fertile, the roots might be less inclined to spread out quickly in order to find nutrients.

When you have back-filled about half way, water the soil so it settles around the roots and continue back-filling until you reach the desired height. Water again thoroughly to make sure there are no spaces around the roots and they all have direct contact with the soil.

Using a rubber tree tie, attach the tree to the stake near the bottom of the trunk so the tree canopy has free movement in the wind. Don't use anything but a proper tree tie to attach the tree to the stake and definitely don't use a piece of rope! As the tree grows, you will be able to loosen the tree tie so you don't cause damage to the trunk.

It is really important you keep your new tree well-watered for the first few months after planting. Spread compost mulch around the base of the tree to preserve soil moisture and help it cope with its first winter.

## With these rings I will tell you many secrets

The science of dendrochronology (studying tree rings) can tell us a lot about events in the life of a tree. Every year, the lateral meristems add another layer of sideways growth. In spring, when the weather is mild and wet, the growth is rapid and therefore the wood appears lighter in colour. In summer, when there is less rainfall, the growth is slower and the wood appears darker. It is this light and dark wood that gives the tree its very distinct rings.

By looking at the rings, you can tell the

years when the tree was at its happiest. Even, wide spacing of the tree's rings show a period of time when rainfall was abundant and the climate was just right for the tree. You can also tell when the tree was not so happy as the rings will be closer together indicating a period of drought. In this way, you can look back into history and know exactly what the weather was like. Scarring on the rings will tell you if the tree has survived a forest fire and you can tell the exact year when that happened. Tree rings also indicate when something might have been leaning against the tree, if it was being shaded out by neighbours for a time and there will also be signs of insect attack.

Trees only record the general climate during their period of growth so a tree's rings won't tell you anything about what happened in winter but they will, of course, give us a fairly accurate record of the age of the tree. If the General Sherman redwood tree really is 2,500 years old, that means he put down his first ring in around 480 BCE and he can tell us a lot about what's happened to him since then.

## Setting the alarm

Few of us are lucky enough to wake up naturally at just the right time we need to so, before we go to sleep we must set our alarm clocks. In my case, I need three alarms to wake me up! When plants go off to sleep, they must wake up at the right time – too early and their new, fresh leaves could be damaged by a sudden frost, too late and there won't be enough time to accumulate sugars sufficient for the energy-sapping task of flowering.

Without an alarm to wake us up and if we didn't have to be at work on time, we would begin to feel sleepy as the hours of darkness approach and we would naturally start to wake up as daylight comes. Even with our eyes shut, we can still sense the coming of the dawn on some level or another. So it is with many plants and they gradually sense the decreasing hours of darkness and increasing hours of daylight even though they are effectively asleep. Thus, they respond to daylight and, even if the weather is unseasonably warm, may not bother getting up until there is enough daylight to make it worth their whiles going back to work.

Many flowering plants have become used to cold winters and they have developed a cell memory of how long those cold winters normally last in the places where they originated. They will not be fooled by a warm spell in winter because they have set their alarms. Just as our alarm clocks tick away in the background, counting the hours we have determined we must sleep for, so these plants count the hours they must sleep for too.

But they don't count all the hours, just the

amount of time the temperature is below a certain value – for some plants this could be zero degrees Celsius, for others it could be higher than that. When the temperature drops to the required level, the alarm clock is set and the plant starts counting only the hours when it is below the temperature it has determined for itself. Only when the plant has accumulated enough hours at that temperature – known as chill units – will its buds wake up and break out of dormancy.

Fruit growers set great store by the amount of time their fruit plants set on their alarm clocks. In areas with normally long, cold winters, they will look for varieties that need a high number of chill units to break dormancy. Some apple trees, for example, have a high chill unit requirement so in areas with warmer winters, these plants flower later and produce fruit later, meaning they may not produce enough fruit to be commercially viable.

Just because we get a warm winter, it doesn't necessarily follow that our flowering plants will perform nice and early. For those plants waiting for that magic number of colder hours, their flowers might be unusually late. The worry is, if the trend for warm winters persists, some of our plants may not flower at all. Of course the wild plants will adapt. Effectively, they will simply move north to a place where winters are just the way they like them. For us, as gardeners, global warming may mean we have to choose a completely different set of plants to grow.

# Life in the Graveyard

When plants die, scores of creatures rush in to take advantage of a sudden supply of food. In the wild, plants fall where they have lived all of their lives and their decomposition takes place amongst the living. For plants that share their space with human gardeners, upon their deaths, they will probably be removed and interred in the compost heap. No plants live in this plant graveyard but the production of compost is so important to the health of our plants, it is vital for a good plants-person to understand what goes on in the average compost heap.

Living organisms break down the material in a compost heap and, for them to thrive, they must have oxygen. Assuming there is a plentiful supply of air in the compost heap, it will always be teeming with life in various forms.

The first life forms to take an interest in dead plant material are the tiny psychrophilic micro-organisms – mainly bacteria. All those sugars and proteins in the newly-dead plants allow the psychrophilic micro-organisms to increase rapidly. They eat as much as they can and, recharged with all that energy, they reproduce extremely quickly. Eventually, they increase in numbers so they become victims of their own success. They can only survive in cooler temperatures and, assuming you have added a lot of plant material to the heap all at once, all their activity causes temperatures in the heap to rise so high, it just gets too hot for them.

Now it is the turn of the mesophilic micro-organisms to pile in. They can survive at slightly higher temperatures than the psychrophilics but they too become so successful at breeding that they increase the temperature to values even they cannot withstand.

When it all gets too hot for the mesophilic organisms, the thermophilic organisms move in. If the increase in temperature is sufficient to support the thermophilics, then you get more rapid decomposition of plant material and, if it gets hot enough for long enough, most seeds, persistent plant roots and plant diseases will be killed off. Encouraged by the heat they love, the thermophilics breed so rapidly they use up most of the available oxygen until, eventually, they die.

After the demise of the thermophilic organisms, the heap begins to cool and the mesophilics can move back in to continue eating the last scraps of plant material. With the amount of available oxygen in shorter supply, their sexual activities are much reduced, they are sluggish and so they don't create quite so much heat as they initially did. At this stage, human intervention in the form of turning the heap to make air available again will initiate a repeat of the whole heating up process and the plant material will break down even further and faster.

Eventually, the plant material breaks down so much as to render it no longer of interest to the bacteria and turning the heap doesn't serve to heat it up. Now the compost is cool enough to allow all manner of creatures to move in and this little plant graveyard becomes, for a time, a whole city of activity.

Fungi become more active as the heap cools down and they begin to finish the decomposition process started by bacteria. Springtails – tiny little insects that jump around much like fleas – are attracted to the heap to feed on fungi, algae and decomposing plant material. Slugs, snails, millipedes and others are attracted to the decomposing plant material as well. Wood lice take advantage of any rotting woody bits and play a part in helping to break these tougher materials down. Worms ingest the organic matter and produce nutrient-rich excretions.

Following right behind the creatures that eat the organic matter are their predators. Spiders prey on springtails and other small invertebrates, while centipedes prey on the spiders as well as everything the spiders might eat. Rove beetles and ground beetles are attracted to the slugs and snails. Toads and slow worms may take refuge in the relatively undisturbed safety of a cooling compost heap and, while they're there, will snack on slugs and snails amongst others.

Once compost has reached a stage where the larger invertebrates are happy to move in, it enters the curing stage where it should remain quite cool for a period of time to allow the final decomposers to do their work. Eventually the pH in the heap settles down closer to neutral and the amount of humus (fully decomposed organic matter) increases. Humus carries a negative charge which is able to attract positively charged nutrients, thus the final stage of composting increases the compost's ability to hold onto nutrients and renders it suitable for use.

## How to make compost in a hot heap

Hot heaps – those turned constantly to maintain a high temperature – are the quickest way to make compost and have the added advantage in that most perennial weed roots, weed seeds and plant diseases are destroyed as well. As with any method of making compost you need to get the ingredients right – or rather the ratio of ingredients.

To make good compost you need a ratio of carbon to nitrogen of around 30 parts carbon to 1 part nitrogen in dry weight. However, no one has the time to dry out their materials and then carefully measure them. To simplify matters, the fifty – fifty volume brown to green materials method works perfectly well.

Browns are carbon-rich, dry materials such as cardboard boxes, paper, woody stems, straw and even natural fibre clothing. They break down more slowly than the wetter stuff and hold their shape for longer. In this way, structure is maintained within the heap and air isn't pushed out.

Greens are soft, wet materials, rich in nitrogen. This includes grass clippings, annual weeds, food waste, tea bags, coffee grounds and vegetarian pet excretions. They break down very quickly and encourage bacteria, insects and fungi to enter the heap.

You can add cooked foods and dairy products to a carefully managed hot heap but these foods will encourage flies, maggots and rats and I would recommend you keep dairy, cooked foods and meat out of even a hot heap.

Once you have the ingredients right, the key to a good hot heap is turning regularly to allow air to enter the heap and force it to heat up again. You also need to accumulate a good amount of materials before you start turning and, once you begin to turn, you shouldn't add any fresh material. In addition, compost created in a hot heap must go through a curing process otherwise, if used too 'fresh', it may do damage to sensitive plants.

The ideal hot composting system should have four bays – preferably wooden structures at around 1 metre cubed each sitting side by side. Your bays should be situated on soil if possible to allow all the organisms responsible for composting to enter and leave via the soil below.

In bay no. 1 you will begin to amass the ingredients you need. Put all your garden waste in there plus waste paper and cardboard until you have filled the bay. Then you turn from bay 1 into bay 2. When you're next feeling energetic, turn the heap from bay 2 into bay 3 while you begin to fill bay 1 again with fresh plant material and then continue turning backwards and forwards between bays 2 and 3. How often you turn depends largely on how fit you feel. The more frequently you turn, the less time it takes for the compost to get to a stage where it is ready for curing. If you only turn once a month or so, it will take longer.

When the heap no longer produces any heat after you turn it, you can put the compost into bay 4 to allow it to cure while you begin the whole process all over again. Leave the compost to cure for at least four weeks. When it's dark and crumbly with a nice, earthy smell, it's ready to use.

# Making compost in a cold heap

A cold compost heap is the most common type. For cold composting you only really need one bin or bay and you can add materials to the top of the heap regularly so you don't need to put aside an area for amassing ingredients. This means you just add the right stuff in the right quantities and pretty much leave it to itself. If it gets too wet and mushy and begins to smell a bit, you haven't got enough air in the mix, usually caused by too many greens. Simply give the heap a stir to put back the air and add some brown material. If the heap isn't doing much at all, you probably have too many browns and the heap has become too dry. In this case, you need to give the heap a good water and add some greens to wake it up again.

Once you've got the mix right, all you need is time to pass (usually more than a year) before you can start removing ready compost from the very bottom of the pile. Because you are only adding small amounts of waste at any time, it is really only the cooler-living psychrophilic micro-organisms and the larger invertebrates that do the work. This is because there will not be enough fresh food at any given time to create the kind of population explosions required to generate heat. In a cold heap, the organisms nibble away slowly and steadily at the food within and this is why it takes a year or more to be ready.

I prefer cold heaps – they require very little input from me and that's the way I like it. I just add my kitchen scraps, my grass clippings and a generous amount of paper and cardboard over the course of a year and, each spring when I want to replenish the organic matter in my soil, there's usually enough well-rotted stuff at the bottom of the compost bin. The only energetic bit is when it comes to harvesting the ready compost from the bottom. This can be a little tricky as you have to slide your spade into the bottom whilst trying not to disturb the fresh material at the top.

A cold heap will never heat up enough to kill the roots of perennial weeds or even the seeds of some annuals and putting perennial weeds in a cold compost pile allows the roots to survive and re-grow when you use the compost later on. However, you can put perennial weeds and weeds with seeds into a cold heap if you pre-treat them first. You need to put them in a black plastic bag, excluding light and air for a good few months until they have turned black and mushy and then you can be pretty much sure they're dead. What you are really doing is suffocating the roots enough so they can't regenerate and the contents of the bag will smell a bit when you open it because of that.

After they're sufficiently mushy and stinky, you can put them on your heap.

If you have a household recycling scheme that allows you to dispose of garden waste separately, you can put perennial weeds in your garden waste bin. Your local council will be able to compost difficult items like persistent weeds and woody prunings within a hot system capable of successfully kill off the roots and breaking down the woody material. You can usually either buy back the resultant compost or sometimes get it for free from council collection points.

## Greens and browns

Once you know which materials are greens and which are browns, you can get the mix right whatever composting system you decide is best for you. Here's a list, though not exhaustive, of some of the more common ingredients:

| Greens | Browns |
| --- | --- |
| Grass clippings | Paper |
| Weeds | Cardboard |
| Tea bags and tea leaves | Hoover contents |
| Vegetable peelings | Pet and human hair |
| Fruit peelings | Wooden bbq skewers |
| Pond debris | Floor sweepings |
| Seaweed | Feathers |

| Hay | Natural fibre clothing |
| Coffee grounds | Pencil sharpenings |
| Straw | Old cut flowers |

A mix of both green and brown

Vegetarian pet bedding

Horse manure with straw bedding

Evergreen prunings

Houseplants

Neither green nor brown

Egg shells add calcium to the finished compost

# What now for this book?

I hope you have enjoyed reading this book and I hope you learned some things about plants you didn't already know.

I am not generally a hoarder – in fact, quite the opposite - but there are just two things I do hoard. At least in my private life, outside of my work, I have great difficulty throwing away healthy plants and sometimes even sickly ones – hence my own garden is not as tidy and healthy as you might think! The other things I hoard are books. For some reason I cannot bring myself to throw away books and I just have to keep building more bookshelves to accommodate them. I hope you will want to keep this book on your bookshelf but, if you must throw it away, I cannot think of a better place to put it than on a compost heap.

To compost this and any other book, newspaper or magazine, tear out the pages one by one and scrunch them up into a ball before you put them in the heap. Scrunched up in a ball, the paper traps much-needed oxygen for your composting organisms and even provides hiding places and temporary homes for some.

# Index

## A

# B

# C

# D

# J

# L

# M

# R

# S

# T

# X

# Y

# If plants could read, they'd read Gardenzine

Gardenzine Publications

***www.gardenzine.co.uk***

Coming soon:

# The Plant City

*By Julie C. Kilpatrick*

What do plants get up to deep down in the soil where we can't see them?

Plants go to work above ground but make their homes in the soil. There, they communicate with their friends, have their dinner, even go to sleep.

Everyone knows a healthy soil means a healthy plant but, because we can't see what goes on there, we often neglect it.

The second book in the Plant Listener series takes you into the world of the soil and why your soil behaves the way it does. It unlocks the secrets of soil nutrition and tells you how, without all that much effort at all, you can create the perfect home for your plants.

Lightning Source UK Ltd.
Milton Keynes UK
UKHW030707080919
349373UK00001B/1/P